The Chinese Treasure

WORLD HERITAGE IN CHINA

Compiled by Luo Zhewen, Li Min

China Architecture & Building Press

Preface

Creams of Nature and Gems of China

Luo Zhewen

World heritage is a masterwork of nature and human civilization, and is the unparalleled natural and humanistic and cultural sceneries. For the purpose of protecting world culture and natural heritage, at the 17th session of UNESCO held on 16th of November, 1972, the Convention Concerning the Protection of the World Cultural and Natural Heritage was officially adopted. In 1976, The UN World Heritage Commission (WHC) was founded. And in 1978, it approved the first group of 12 world heritages to be included on the World Heritage List.

World heritage includes natural heritage, cultural heritage, cultural landscape, and mixed cultural and natural heritage. Any place when it is named on the list of world heritage by the WHC, will become a world-class scenic spot and will be supported by the World Heritage Fund, and be widely noticed and strictly protected by the world community. At the same time, its reputation will also be greatly promoted and a decent social and economic result will be produced.

China is one of the world's four old civilized countries, enjoying a vast territory and beautiful mountains and waters. Stretching across thousands of miles and having a history of five thousand years, China's magnificent and gorgeous natural landscapes and long history have carried in itself the nation's brilliant cultural and natural heritages, a valuable wealth also shared by the whole world. On 12th of December of 1985, China joined the Convention Concerning the Protection of the World Cultural and Natural Heritage and on October 29 of 1999, was elected as a member of the WHC. Over the past 25 years, Chinese government together with its people have spared no efforts to protect and utilize world heritages and have made distinguished contributions to this project.

Generally speaking, China's world heritage sites have five essential features:

1. A Great Number and Wide Distribution

In 1986, China submitted its first application to UNESCO for the nomination of its projects on the world heritage list. At the 11th session of the WHC in 1987, China's six projects including the Imperial Palace were approved to be included on the list of world heritage. Until June, 2009, a total of 38 cultural and natural scenic spots had been listed on the world heritage list, including seven natural heritage, 27 cultural heritage and four mixed cultural and natural heritage. They are distributed in 27 provinces, autonomous regions and municipalities directly under the central government and special administrative regions. In addition, the Chinese *kunqu* opera, ancient *guqin* art and the muqam art of Uyghur, and the long tune folk songs of Mongolians have been listed as the verbal and intangible cultural heritages.

2. Splendid and Diverse Landscapes

China has a land territory covering an area of about 9.6 million km^2. It has the highest peak of Mount Qomolangma, 8,844 m above sea level. China's territorial waters is about 4.7 million km^2. including 7,600 large and small islands. Its eastern and southern coastlines extend 18,000 km long. China's natural landscapes are magnificent and gorgeous, and diverse in types and styles. So far, some of them have been included on the list of world natural heritage such as the magic landscape of Wulingyuan, the fairyland of Huanglong and Jiuzhaigou, the imposing Three Parallel Rivers , the dreamy Karst in South China, the unique gorgeous Mount Sanqingshan landscape and the sanctuaries of giant pandas with exuberant growth of bamboos. They all demonstrate the abundant resources of sceneries and bio-diversity in China.

3. A Long History with a Brilliant Culture

China is a unified state with multiple ethnic groups, comprising of 56 nationalities. Its civilization has been developing into today through a long period of more than 5,000 years, known as one of the oldest and longest civilizations in the world. The images of "dragon", "phoenix", "*kylin*"(Chinese unicorn), and "turtle", are called the "four spirits." They are the totems worshipped in ancient China. The Chinese Han language and written characters are one of the languages and written words used by most people and prevalent in the widest area in the world. From the Great Wall meandering wave upon wave, to the Potala Palace on the roof of the world, and from the Lijiang old town surrounded by water of melted snow to exquisite and elegant Suzhou gardens, the 27 cultural heritage sites on the list of world heritage present a full picture of the long history and brilliant culture of the Chinese civilization.

4. Landscape and Humanity Contrast Finely with Each Other

Upholding the ideal of "unity of Heaven and man" is the traditional thought of Chinese philosophy and aesthetic judgment. The combination of culture and nature is the basic idea of Chinese civilization. "The land is so rich in beauty that has caused unnumbered heroes to bow in homage," a line of the verses written by the great leader Mao Zedong, in which he summarized the historical process of the glorious contrast between landscape and humanity. From Mount Taishan, first of the Five Sacred Mountains to Mount Huangshan which is always surrounded by mists and clouds, and from the sacred Mount Emei and the Leshan Giant Buddha, to the Mount Wuyi which is famous for the origin of the Neo-Confucianism in Tang and Song period and the danxia landform, China's mixed cultural and natural heritages have opened an everlasting scientific window for the whole world on the knowledge of China.

5. Protecting and Making Use of Heritages Forever

China has a long tradition of attaching importance to natural heritages as well as humanity and history. The Confucianism advocated by Confucius has been passed down to nowadays and is still popular. In the drive to transform China gradually into a modernized country, Chinese government and its people will pay high attention to the protection of its best natural and cultural heritages. They will spare no efforts to carefully protect and make good use of the country's world heritages. In order to promote the whole nation's consciousness and action for the protection of heritages, the State Council made a decision that starting from the year 2006, the second Saturday of June each year is designated as the "China's Cultural Heritage Day." China as a large country of world heritages is undergoing a dauntless and outstanding task for inheriting and passing down the human civilization and protecting the natural resources of the earth.

So many gems are produced in China. So beautiful are its landscapes. This book has given a full description of and introduction to the China's world heritages. We have tried our best to make it more vivid and easily readable through its illustrations and narratives. It is the newest edition of its kind in the country. We present this book as a gift to the 60th anniversary of the birthday of the People's Republic of China.

CONTENT

World Cultural Heritage

The Great Wall

China's Great Wall was a defense project in ancient times. It took the longest ever time to build and made a record volume of work. Beginning from the 7th to the 8th century B.C., the construction had lasted for more than 2,000 years without break. It was built in the north and central China, extending more than 50,000 km long, known as a gigantic construction in the world, scarcely seen "in the past 2,000 years over an immense area." We can make an estimate of the amount of work in the construction of the 7300-km-long Great Wall during the Ming Dynasty that needed an amount of brick-and-stone work of 50 million m^3 and earthwork of 150 million m^3. If we built a road of 10 m wide and 35 cm thick with these stones and earth, it can stretch around the globe more than two rounds in circumference. So, hundreds of years ago, China's Great Wall was known in parallel with the Wrestling Ring in Rome and the Leaning Tower of Pisa in Italy as one of the Seven Wonders in the Middle Age.

The Great Wall starts from Shanhaiguan Pass in the east and ends at Jiayuguan Pass in the west. Traversing through all kinds of complicated topographical forms of mountains and ridges, deserts and grasslands, precipitous cliffs and deep rivers and with its unique wall structure, it was really a wonder in the history of the ancient Chinese constructions. It was not a simple wall construction, but was a multi-functional and complete military defensive system with various constructions of walls, towers on battlements, passes of wall, pier-based strongholds, barracks, guardhouses, garrison posts, and beacons, controlled by military commanding headquarters at all levels. According to historical documents, a total of one million military troops had been dispatched to guard along the Great Wall during the Ming Dynasty.

Through the ages, so many noted historical "dramas" had been staged around the Great Wall, in which a number of outstanding strategists and politicians came to the fore. Many great and soul-stirring campaigns occurred at the foot of the Great Wall and many historic incidents of dynastic changes would have had something to do with whether the Great Wall was well guarded or not. All these historical facts have brought about the ever-growing cultural and historical significance of this ancient construction. In the hearts of the Chinese people, the Great Wall is like a great dragon meandering across the whole territory of China.

In 1987, UNESCO included China's Great Wall on the list of World Cultural Heritage and an evaluation made by the World Heritage Commission is as follows:

In c. 220 B.C., under Qin Shi Huang, sections of earlier fortifications were joined together to form a united defence system against invasions from the north. Construction continued up to the Ming Dynasty (1368–1644), when the Great Wall became the world's largest military structure. Its historic and strategic importance is matched only by its architectural significance.

2

3

_1 Jinshanling section of the Great Wall
_2 Badaling section of the Great Wall
_3 Tower of the Shanhaiguan Pass

_4 Mutianyu section of the Great Wall

5

_5 Jiayuguan Pass
_6 Laolongtou of Shanhaiguan Pass
_7 Laolongtou of Shanhaiguan Pass

6

7

_8 Jiankou section of the Great Wall
_9 Moonlight night at the Jiayuguan Pass
_10 Jinshanling section of the Great Wall

Imperial Palaces of the Ming and Qing Dynasties in Beijing and Shenyang

Imperial Palace in Beijing, also known as "Forbidden City", is located in the center of the Beijing City, as the residence of rulers of the Ming and Qing dynasties. It was first built by Zhu Di, Emperor of the Ming Dynasty, and after 14 years the construction was completed (1407-1420). The Palace is a plane rectangular in shape, 961 m long from north to south and 753 m wide from east to west, stretching over about 720,000 m^2. It is enclosed by walls on four sides in circumference of 3,428 m. The wall is 7.9 m high on a wall base of 8.62 m wide. Its upper width is 6.66 m. On the outward side of the upper part are crenellations, and eaves were built on the inward side. An exquisite corner tower (turret) was built on each of the four corners of the palace. There is a city moat around the walls, 52 m wide, and 3,800 m in circumference, completing its whole defensive system. There are four gates on the four sides of the walls. On the front of the south wall is the Meridian Gate. And on the north is the Gate of Divine Prowess; east is the East Prosperity Gate and west the West Prosperity Gate.

In a period of 491 years from 1420 to 1911, a total of 24 emperors of the Ming and Qing dynasties had lived and ruled the country within this imperial palace. In the palace, there are more than 9,000 halls and rooms of various kinds with a total area of 150,000 m^2. The Forbidden City is composed of the front and back sections, or called outer court and inner court. The front section contains the Hall of Supreme Harmony, Hall of Central Harmony and Hall of Preserving Harmony as its centre and on the east and west sides of the central axial alignment, there are Hall of Literary Brilliance and Hall of Martial Valor as its wings, where the emperors dealt with their government affairs and held ceremonies for important events. The back section takes the Palace of Heavenly Purity, the Hall of Union and the Palace of Earthly Tranquility as its center, on both sides of which there are Six Eastern Palaces and Six Western Palaces, in addition to the Hall of Mental Cultivation, Hall of Ancestral Worship, Hall of Abstinence, Hall for the Sound Growth of the Successor, Palace of Tranquil Longevity, Hall for the Consolation of Mothers and the Imperial Garden which were the residence of emperors together with their family members and the place for worshipping Buddha and study and entertainment. The general layout of the Imperial Palace was spread according to a central axial alignment with symmetric buildings on its both sides. The front three halls and the back three palaces are set on the central axial alignment. These halls and palaces were all magnificently and elegantly constructed, displaying the highest skill level of ancient Chinese architecture and becoming the largest and best-preserved ancient imperial palace in the world. Its architectural pattern is really an incomparable masterpiece. In 1961, the State Council announced that the Imperial Palace in Beijing was one among the first batch of the key cultural relics under state protection.

The Shenyang Imperial Palace was first built in 1625 by Nurhaci, first emperor of Qing Dynasty in northeast China and his son Abahai, also known as Emperor Taizong of Qing. It is also called Imperial Palace of Prosperous Capital, and after Qing Dynasty moved its capital into

Beijing, the Shenyang Imperial Palace was also changed its role to become an auxiliary palace. It has now 114 ancient buildings stretching over 60,000 m^2. The palace buildings can be divided into three sections according to their layout and construction time, the eastern route involves the Great Administration Hall and Ten Princes Pavilions, built during the period of the reign of Emperor Nurharci; the central route built during the period of the reign of Emperor Taizong includes Grand Qing Gate, Holy Administration

1

2

_1 Looking down at the Beijing Imperial Palace
_2 Meridian Gate of the Beijing Imperial Palace
_3 Looking down at the Beijing Imperial Palace

4

Hall, Phoenix Tower, and Palace of Peace and Tranquility, Hall of Turtle-dove, Hall of Celebration and Hall of Happiness Sign; the western route includes the Book Source Pavilion and others built during the period of the reign of the Qing Emperor Qianlong. A great number of buildings scattered in the whole court of the palace, present to view their elegance and majesty.

In 1987, UNESCO included the Beijing Imperial Palace on the list of World Cultural Heritage. In July 2004, at the 28th session of the WHC, the Shenyang Imperial Palace was approved as the extended part of the Ming and Qing Imperial Palace to be included on the list of World Heritage. The evaluation made by the WHC is:

Seat of supreme power for over five centuries (1416-1911), the Forbidden City in Beijing, with its landscaped gardens and many buildings (whose nearly 10,000 rooms contain furniture and works of art), constitutes a priceless testimony to Chinese civilization during the Ming and Qing dynasties. The Imperial Palace of the Qing Dynasty in Shenyang consists of 114 buildings constructed between 1625–26 and 1783. It contains an important library and testifies to the foundation of the last dynasty that ruled China, before it expanded its power to the centre of the country and moved the capital to Beijing. This palace then became auxiliary to the Imperial Palace in Beijing. This remarkable architectural edifice offers important historical testimony to the history of the Qing Dynasty and to the cultural traditions of the Manchu and other tribes in the north of China.

_4 The Hall of Supreme Harmony in Beijing Imperial Palace

_5 The Nine-dragon Wall in Beijing Imperial Palace

_6 Bronze Lion in front of the Hall of Supreme Harmony in Beijing Imperial Palace

_7 Inside view of the Hall of Union in Beijing Imperial Palace
_8 Inside view of the Hall of Supreme Harmony in Beijing Imperial Palace

8

弘敷五典無輕民事惟難
乾隆丁巳恭摹
皇祖御書
惟精惟一道積于厥
明

9

_9 Inside view of the Palace of Heavenly Purity in Beijing Imperial Palace

11

_10 The Imperial Garden in Beijing Imperial Palace
_11 East Chamber of Warmth in the Hall of Mental Cultivation in Beijing Imperial Palace

日升月恒
囍

_12 A bridal chamber in the Palace of Earthly Tranquility for the Qing emperors in Beijing Imperial Palace

14

_13 Stone sculpture in Beijing Imperial Palace
_14 Inside view of the Study of Pure Fragrance in Beijing Imperial Palace
_15 Inside view of the Hall of Central Harmony in Beijing Imperial Palace
_16 A night scene of the Turret of Beijing Imperial Palace

15

_17 Great Administration Hall at Shenyang Imperial Palace
_18 Holy Administration Hall at Shenyang Imperial Palace
_19 Details of the Holy Administration Hall at Shenyang Imperial Palace
_20 Phoenix Tower at Shenyang Imperial Palace

17

18

19

Mausoleum of the First Qin Emperor

Established in the period between 246—208 B.C., the Mausoleum of the first Qin Emperor is situated at the Mt. Lishan of Lintong County, 30 km east of Xi'an City, Shaanxi Province, and was the mausoleum of the first emperor of China Qin Shihuang. The existing tomb is 76 m high. The whole layout of the mausoleum imitated the pattern of the capital of Xianyang of Qin Dynasty. It was composed of the interior and exterior sections and the interior section is 2.5 km in circumference and that of the exterior section is 6.3 km. The tomb is located at the southwest of the interior section facing the east, where were laid the inner and outer coffins and many other artifacts buried together with the demised. It was the centre of the constructions in the mausoleum, and has been kept intact without being unearthed.

The mausoleum is one of the monarch's mausoleums in the world which has the largest constructions, is unique in structure, and richest in inner contents, becoming a sumptuous underground palace. According to historical documents, Qin Shihuang began building this mausoleum at the age of 13 when he first ascended the throne. His chancellor-in-chief Li Si planned and designed the project, and his general-in-chief Zhang Han held the position of supervisor. The whole construction project took a long period of 38 years. At that time, the total population of the Qin Dynasty was about 20 million, while more than 720,000 laborers were forced to build the mausoleum. It set the first example of the most extravagant mausoleum ever built in the history of Chinese emperors.

The terracotta warriors were buried in some pits with the demised Qin Shihuang at a place 1,500 m away east of the mausoleum. For the time being, three of the pits have been found located in the west and facing to the east, distributed in shape of a triangle. About 7,000 terracotta warriors of Qin together with a great number of modeled horses, chariots and weapons have been excavated. The warriors were made stalwartly with an average height of 1.8 m in diverse and lively appearances, which demonstrated the high techniques of sculpture and modeling industry at that time. The modeled battle array of warriors displayed once again the prowess of the Qin military forces in the unification of the whole Chinese territory. The discovery of the terracotta warriors was a significant event in the history of the world archeology, being honored as the eighth wonder of the world, parallel in popularity with the well known Pyramid in Egypt and the sculptures in ancient Greece.

In 1961, the State Council designated the mausoleum as one in the first group of the key cultural relics under state protection. In 1987, UNESCO approved the inclusion of the mausoleum on the list of World Heritage. The evaluation made by the World Heritage Commission is as follows:

No doubt thousands of statues still remain to be unearthed at this archaeological site, which was not discovered until 1974. Qin (d. 210 B.C.), the first unifier of China, is buried, surrounded by the famous terracotta warriors, at the centre of a complex designed to mirror the urban plan of the capital, Xianyang. The small figures are all different; with their horses, chariots and weapons, they are masterpieces of realism and also of great historical interest.

2

_1 Terracotta warriors of Qin Dynasty
_2 Terracotta warriors of Qin Dynasty
_3 Terracotta warriors and horses of Qin Dynasty

_4 No.1 pit of terracotta warriors and horses of Qin Dynasty
_5 Terracotta warriors and horses of Qin Dynasty
_6 Terracotta warriors and horses of Qin Dynasty
_7 Terracotta warriors and horses of Qin Dynasty
_8 Terracotta warriors and horses of Qin Dynasty
_9 Terracotta warriors and horses of Qin Dynasty
_10 Terracotta warriors and horses of Qin Dynasty –Soldier driving a chariot

5

6

4

7

8

_11 Bronze chariot and horses in a pit of terracotta warriors
_12 Terracotta warriors and horses of Qin Dynasty
_13 Terracotta warriors and horses of Qin Dynasty
_14 No.1 pit of terracotta warriors and horses of Qin Dynasty

三号坑 Pit No.3

Mogao Caves

Located at the Hexi Corridor in Gansu Province, The Mogao Caves are well known for its exquisite and fine Buddhist frescos and statues. They were first built during Qianqin Dynasty in the period of the Sixteen States, and later continued in the Northern Dynasties, Sui Dynasty, Tang Dynasty, the Five Dynasties, Western Xia Dynasty, and the Yuan Dynasty, gradually developing into a gigantic construction complex. For the time being, there are 735 caves with frescos and murals distributed in an area of 45,000 m^2. of which 2,415 painted sculptures were made of clay. It is the largest and richest Buddhist sanctuary of fine arts in the world. In modern times, a library cave where more than 50,000 historical relics had been kept was discovered, from which a science of Dunhuang Studies has been derived—specializing in the research of the ancient documents in the library cave and the fine arts of Dunhuang.

Murals of the Mogao Caves, painted on walls and ceilings of the caves and within the shrines, involved seven motifs such as the Buddhist portraits, Buddhist stories and historical relics, changes of scriptures, gods and spirits, sacrifice donators, and other decorative patterns, as well as other pictures describing hunting, tillage, weaving, transportation, warfare, constructions, dancing performance, marriage and funeral arrangements, in relation to various aspects of the then social life. They were vigorously or beautifully drawn, showing the styles and features of the fine arts in different stages.

Mogao Caves were cut as a large cave temple combining mainly painting, sculptures and architectural art as a whole. The types and forms of the caves include meditation caves, caves with a central tower, hall caves, caves with a central Buddhist altar, caves with three shrines enclosed by walls on four sides, caves of large portraits and Nirvana caves. They were made in different sizes, and the largest was the cave No. 16, occupying a space of 268 m^2. The smallest cave No.37 was less than one-third meter in height. Outside the caves, there had been some wooden halls linked by passages or trestles, but now most of them no longer exist.

In 1961, Mogao Caves were designated by the State Council as one in the first group of the key cultural relics under state protection. In 1987, they were included by UNESCO on the list of World Cultural Heritage. The evaluation made by the World Heritage Commission is:

Situated at a strategic point along the Silk Route, at the crossroads of trade as well as religious, cultural and intellectual influences, the 492 cells and cave sanctuaries in Mogao are famous for their statues and wall paintings, spanning 1,000 years of Buddhist art.

_1 A full view of the Mogao Caves

1

_2 A temple of Mogao Caves
_3 A statue of Buddha in Mogao Caves
_4 A colorful painting in Mogao Caves

4

_5 A statue of Buddha in Mogao Caves
_6 A colorful painting in Mogao Caves
_7 A colorful painting on the popular teaching of Amita sutra in Mogao Caves
_8 A colorful painting in Mogao Caves

8

9

_9 A fresco of Mogao Caves
_10 The nirvana-form of Buddha in Mogao Caves
_11 The nirvana-form of Buddha in Mogao Caves

10

Peking Man Site at Zhoukoudian

The relics of Peking Man of the Old Stone Age were discovered at the Longgu Hill of Zhoukoudian Village, Fangshan District, southwest of Beijing City. The hill was composed of a series of natural caves. In 1923, J. G. Anderson, a Swedish scholar, discovered some fossils of human teeth on the hill. In 1927, a Canadian scholar Davidson Black began excavating the Zhoukoudian site and officially named the three fossils of human teeth found in the village as "Peking species of Chinese Ape Man." On the 2nd of December 1929, Chinese scholar Pei Wenzhong discovered the first Peking Man's whole skullcap fossil in a cave extending 140 m long from east to west. That discovery made a sensation throughout the world. In 1936, under the auspices of Mr. Jia Lanpo, some archeologists discovered three more skullcaps. Up to 1937, enormous human bone fossils of about 40 individual humans had been unearthed from the village, which provided evidences to the presumption that the Peking Man had lived in the Zhoukoudian area in 500,000 to 200,000 years ago.

Also discovered in the site were some primitive stone tools and some vestiges of using fire, making the time of fire using

by mankind in history to be advanced by several hundred thousand years. On the site, five layers of ashes, three heaps of ashes of burned substances together with a great amount of burned bones were found. The thickest of the heaps of ashes was up to 6 m high. All these testify that Peking Man had known how to make use of fire and keep the kindling and known to select some rocks or stones to make weapons or production implements. Peking Man was a species of protohominid in the middle, as a link, of the development from the paleopithecus (fossil anthropolid) to the homo sapicns (man). The discovery has produced a significant impact on the research of biology and the history of human development.

In 1961, the Peking Man site was designated by the State Council as one in the first group of the key cultural relics under state protection. In 1987, it was listed by UNESCO as one of the sites of World Cultural Heritage. The evaluation made by the WHC is:

Scientific work at the site, which lies 42 km south-west of Beijing, is still underway. So far, it has led to the discovery of the remains of Sinanthropus pekinensis, who lived in the Middle Pleistocene, along with various objects, and remains of Homo sapiens dating as far back as 18,000–11,000 B.C. The site is not only an exceptional reminder of the prehistorical human societies of the Asian continent, but also illustrates the process of evolution.

2

3

_1 The ruins of Zhoukoudian
_2 The ruins of Zhoukoudian
_3 The museum of Zhoukoudian ruins
_4 The ruins of Zhoukoudian

Historic Ensemble of the Potala Palace, Lhasa

Potala is a word transliterated from the Sanskrit "Potalaka," originally meaning the island where the Avalokitesvara Bodhisattva lived. Situated on the Red Mountain in the downtown of Lhasa City of Tibet, the Potala Palace was first built in the 7th century during the reign of the Songtsen Gampo, the 32nd King of the Tubo Dynasty, originally called "Palace of Red Mountain." During the 17th century, the Fifth Dalai Lama rebuilt the majestic palace on the same site of Red Mountain, and renamed it as Potala Palace which had been the government center of the local rulers of Tibet, integrating religion with politics.

The Palace presents a gigantic and magnificent appeal, stretching out along the alignment of the contours of the mountain, covering about 360,000 m^2, with a 13-storey principal construction of 117 m in height. It is not only famous for its Tibetan architectural style, but also a masterpiece of ancient Tibetan palace-style building, honored as "a jewel shining on the roof of the world."

The Potala Palace complex contains the White Palace, the Red Palace, and other supporting buildings around them. The group of constructions tak-

ing the White Palace as its centre was completed in 1648. The White Palace, a 7-storey building facing the east, was the headquarters for Dalai Lamas of the past centuries to handle their daily political and religious routines. The Red Palace, a six-storey building at the center of the group, was completed in 1694. It was a place to worship the shrines (stupas) of all the passed Dalai Lamas and conduct the religious ritual activities. The Potala Palace constructions display the outstanding achievements of the Tibetan, Han, Mongolian and other ethnic groups in their respective cultures, art and religions. Their lofty appearance and the holy status in the Tibetan Buddhism have become the cultural symbols of the Tibetan ethnic group accepted by the whole world.

The Jokhang Temple is situated in the southeast Lhasa City, first built in the year 601 A.C. It was the oldest wood-framed construction in Tibet imitating the Tang style and combining the Han and Tibetan techniques. The eminent monk Tsong-kha-pa founded the Dge-lugs Sect of the Tibetan Buddhism at the beginning of the 15th century. And since then, more and more worshippers had frequented to the temples of this Sect. During the 17th century, the Fifth Dalai Lama further expanded and repaired the Jokhang Temple on a large scale, developing it into a huge construction complex stretching over 25,100 m^2. The chief building in the temple is the four-storey Hall of Sutras-chanting, constructed in the Han style, but with decorations on the tops of columns and eaves in typical Tibetan style. On the ground floor was enshrined a golden statue of the Saint Sakyamuni which had been carried into Tibet by Princess Wencheng, providing historical evidence to the friendly exchanges between the Tibetan and the Han people.

The Norbulingka, or "treasured garden" in Tibetan, is situated at the bank

2

3

_1 The Potala Palace
_2 Inside view of the White Palace of the Potala Palace
_3 The White Palace of the Potala Palace
_4 Statues of Songtsen Gampo, Princess Wencheng and Princess Chizun

of Lhasa River about 2 km west of Potala Palace, covering an area of about 3.60 million m^2. It was first built during the middle of the 18th century, and had been used as a summer palace for Dalai Lamas of preceding generations to deal with administrative routines and conduct religious ceremonies. Some 100 species of ornamental plants were cultivated in the park, including some exotic flowers and rare herbs collected from the south and north of the Himalayas or transplanted from inland or introduced from abroad.

The Potala Palace was listed as one of the sites of World Cultural Heritage by UNESCO in 1994. In 2000 and 2001, the Jokhang Temple and Lorbulinka were also included on the list. The evaluation made by the World Heritage Commission is:

The Potala Palace, winter palace of the Dalai Lama since the 7th century, symbolizes Tibetan Buddhism and its central role in the traditional administration of Tibet. The complex, comprising the White and Red Palaces with their ancillary buildings, is built on Red Mountain in the centre of Lhasa Valley, at an altitude of 3,700m. Also founded in the 7th century, the Jokhang Temple Monastery is an exceptional Buddhist religious complex. Norbulingka, the Dalai Lama's former summer palace, constructed in the 18th century, is a masterpiece of Tibetan art. The beauty and originality of the architecture of these three sites, their rich ornamentation and harmonious integration in a striking landscape, add to their historic and religious interest.

6

7

_5 The golden roof of the Jokhang Temple
_6 Outside view of the Jokhang Temple
_7 The waterside pavilion in the Norbulingka
_8 Decorations on the golden roof of the Jokhang Temple
_9 Inside view of the Jokhang Temple

_10 Details of the golden roof of the Jokhang Temple
_11 Details of the golden roof of the Jokhang Temple
_12 Details of the golden roof of the Jokhang Temple

13

_13 The Padmasambhava in the Jokhang Temple
_14 A corner of the Potala Palace
_15 A distant view of the Potala Palace

14

Mountain Resort and its Outlying Temples, Chengde

The Mountain Resort, also known as an "auxiliary palace in Rehe," is located in the northern part of Chengde in Hebei Province. It seats on a highland circled by mountains, enjoying a salubrious climate. It had been used as a summer resort for emperors of the Qing Dynasty to spend the hot season while dealing with their government affairs.

First built in 1703, or the 42nd year of the reign of Emperor Kangxi of Qing Dynasty, the construction of the Mountain Resort was completed in 1790, or the 55th year of the reign of Qing Emperor Qianlong, stretching over 564 hm^2, including more than 110 buildings. It is the largest imperial garden still existing in China. Lying against mountains and facing to lakes, the summer resort nestles in exuberant vegetations with randomly dotted pavilions or waterside passages, blending the best gardening skills and workmanship of the South and North China as a whole. It presents to view a beautiful landscape well-distributed with limpid waters and green mountains, sequestered pools and isles, sending forth a lively wild interest of nature.

The front gate of the summer resort is

called Righteousness Gate (*Lizheng men*), and the stone walls with crenellations on them imitated the styles and forms of that of the imperial residence in Beijing. They stretch up and down along the slopes of the mountains in a distance of more than 10 km, known as "Mini Great Wall." The summer resort is composed of two sections: the palace section and the garden section. The garden section involves three scenic areas: The area of lakes and isles, area of plain and area of mountains. The garden section is dotted with 36 scenic spots with each a inscription of a four-character name inscribed by Emperor Kangxi, and another 36 scenic spots with each a inscription of a three-character name inscribed by Emperor Qianlong. Qianlong had highly appraised the resort as an interesting place created by nature and more beautiful than those palaces and gardens built by emperors of the Han and Tang dynasties.

A total of 12 majestic temples were built one by one on the hilly ridges around the resort from north to east, just like numerous stars centering on the moon. With different architectural styles, all the temples

2

3

4

5

_1 The Lake Center Pavilions at Chengde Mountain Resort
_2 The Lion Grove Garden at Chengde Mountain Resort
_3 The Lion Grove Garden at Chengde Mountain Resort
_4 Hall of Simplicity and Sincerity at Chengde Mountain Resort
_5 Inside view of a chamber at Chengde Mountain Resort
_6 The Misty Rain Tower at Chengde Mountain Resort

were built by the Qing government in a bid to testify its good will to unite the Tibetan, Mongolian and Uygur ethnic groups, eight of which were directly administered by the government, known as "Eight Temples outside Beijing." Their architectural styles can be classified as Tibetan style, Han national style, and style of blending Tibetan and Han national elements.

In 1994, the mountain resort and its outlying temples were listed as one of the sites of World Cultural Heritage by UNESCO. The evaluation made by the World Heritage Commission is:

The Mountain Resort (the Qing Dynasty's summer palace), in Hebei Province, was built between 1703 and 1792. It is a vast complex of palaces and administrative and ceremonial buildings. Temples of various architectural styles and imperial gardens blend harmoniously into a landscape of lakes, pastureland and forests. In addition to its aesthetic interest, the Mountain Resort is a rare historic vestige of the final development of feudal society in China.

8

_7 The Pavilion among Clouds and Mountains at Chengde Mountain Resort
_8 Inside view of the Hall of Refreshing Mist
_9 The lake scenery of Chengde Mountain Resort

_10 The Pavilion of Mahayana in the Temple of Universal Peace
_11 The statue of Avalokitesvara with a thousand arms and eyes in the Pavilion of Mahayana

_12 The Five-pagoda Gate of the Temple of Sumeru Happiness and Longevity
_13 The pagoda at Eternal Blessing Temple

14

_14 Outside view of the Eternal Happiness Temple
_15 Inside view of the Eternal Happiness Temple
_16 A full view of the Temple of Sumeru Happiness and Longevity

Temple and Cemetery of Confucius and the Kong Family Mansion in Qufu

The sage Confucius in ancient China had been worshipped and honored as "the greatest sage and teacher" across the country. The Temple, Cemetery and the Family Mansion of Confucius, since their establishment, have been regarded as the sanctuaries to commemorate him and promote his doctrine, the Confucianism. They are well known for abundant accumulation of cultural relics and treasures as witness to the long history of traditional Chinese culture, science and art.

Founded in the 478 B.C., the Temple of Confucius is situated at the native town of Confucius in Qufu. It is the largest sanctuary in the country with magnificent and splendid constructions, serving as the site for holding memorial ceremonies to Confucius. It is also a model for about 2,000 temples of Confucius all over the world. Covering a floor-space of 140,000 m^2, the Temple was laid out in nine rows of courtyards along a central axial alignment from north to south, distributed on the left, central and right routes in parallel, 630 m long and 140 m wide, with a total of some 460 rooms, halls and altars and pavilions, 54 gateways, and 13 imperial tablet pavilions. The main hall is the Hall of Great Achievements, huge and gorgeous, being the top of all the halls of Confucius in the world. There are 10 columns engraved with aspiring patterns of dragons in front of the hall under the eaves. Through the past 2,400 years, the Temple has maintained the customs of memorial ceremonies and is the longest-

1

_1 A Portrait of Confucius
_2 The Hall of Great Achievements in the Temple of Confucius

lasting temple and one of the most famous ancient construction complex in China.

Confucius Woods (Cemetery of Confucius) is also called Woods of Greatest Sage, being the graveyard of Confucius and his descendants, stretching over 200 hm^2. In the graveyard, there are ancient trees, under which there are many tombs with stone tablets and lines of stone sculptures. Of these tombs, the graves of Confucius, Kong Li, and Kong Ji, in a line of three generations, were enclosed by red walls. The Confucius Woods have survived for a long period of 2,500 years, becoming the longest, best preserved and the largest family graveyard in the world.

The Kong Family Mansion was the living quarters of Confucius's hereditary descendants in direct line, called "Duke of Developing Sage," making up an area of 164,000 m^2. It was the second largest family living quarters in China to those imperial residences of the Ming and Qing dynasties. It has 463 buildings of various types such as halls, offices, passageways and verandahs with windows laid out in nine courtyards which were distributed parallel on three routes. The eastern route contains the family temples, the western route home institutes, the central route office buildings, and its back section was used as residence. As the oldest noble family in China, Confucius's grandsons in direct line had maintained their titles of nobility for more than 2,100 years. The mansion is the existing largest, best-preserved and most typical ancient construction complex integrating government offices with living quarters as one in China. Its back garden called Iron Hill Garden, quiet and elegant, was laid out with an ingenious design, had been a place often frequented by the "Duke of Developing Sage" of past generations and their family members.

2

萬世師表
斯文在茲

_3 Inside view of the Hall of Great Achievements in the Temple of Confucius
_4 A stone memorial archway in the Temple of Confucius

In 1994, the Temples and Cemetery of Confucius and Kong Family Mansion were listed as one of the sites of the World Cultural Heritage by UNESCO. The evaluation made by the World Heritage Commission is:

The temple, cemetery and family mansion of Confucius, the great philosopher, politician and educator of the 6th–5th centuries B.C., are located at Qufu, in Shandong Province. Built to commemorate him in 478 B.C., the temple has been destroyed and reconstructed over the centuries; today it comprises more than 100 buildings. The cemetery contains Confucius' tomb and the remains of more than 100,000 of his descendants. The small house of the Kong family developed into a gigantic aristocratic residence, of which 152 buildings remain. The Qufu complex of monuments has retained its outstanding artistic and historic character due to the devotion of successive Chinese emperors over more than 2,000 years.

_5 The compound courtyard of the Kong Family Mansion
_6 The high wall of the Temple of Confucius
_7 Inside view of the Hall of Great Achievements in the Temple of Confucius

萬世師表
斯文在茲
氣備四時與天地鬼神日月合其
教垂萬世繼堯舜禹湯文武作之

_8 The tomb of Confucius in the Confucius woods
_9 Stone sculpture in Confucius woods
_10 Stone sculptures in Confucius woods and a tomb
_11 The side hall of the Hall of Great Achievements in the Temple of Confucius

先賢公西子赤之神位
先賢漆雕子哆之神位

Ancient Building Complex in the Wudang Mountains

The Wudang Mountains, also known as the Mountains of Supreme Harmony, are situated in the southwestern part of Danjiangkou City of Hubei Province. During the Ming Dynasty (1368-1644), the Wudang Mountains were named by emperors' edicts as "great mountains" or "profound mountains," enjoying a status higher than those of the Five Sacred Mountains in China. The principal peak of the mountains is called Heavenly Pillar Peak, 1,612.1 m above sea level. Around it there scattered many other magnificent and gorgeous scenic spots including 72 peaks, 36 rocks and 24 ravines.

The ancient building complex of the Wudang Mountains were first built during the Zhenguan period of the Tang Dynasty (627-649) and developed into its prime stage during the Ming Dynasty. In the 31st year of the reign of the Ming Emperor Jiajing (1552), the archway of "Profound Mountains Dominate the World" was established to form a large complex with the "eight palaces and two monasteries" as its center. Today, the ancient building complex mainly include the Palace of Supreme Harmony, Palace of South Rocks, Purple Cloud Palace and Immortal-encountering Palace, and the ruins of Jade Empty Palace and Five Dragons Palace as well as many other types of temples and Buddhist convents, memorial temples totaling more than 200, stretching over more than 1,000,000 m^2. with a floor-space of buildings of 50,000 m^2. The Palace of Supreme Harmony is located on the south side of the Heavenly Pillar Peak, consisting of over 20 buildings such as the Forbidden City, Ancient Bronze Hall, and Golden Hall. The South Rocks Palace is situated under the Solitary Sun Crag of the Wudang Mountains, first built in 1285, and 21 buildings are preserved intact including the stone hall of the Immortal-celebrating Palace, and the Hall of Dragon and Tiger. The Purple Cloud Palace, the largest and best-preserved ancient Taoist building complex in the mountains, is located under the Flag-spreading Peak southeast of the Wudang Mountains, first built in the Emperor Xuanhe period of the Northern Song Dynasty (1119-1125) and expanded in the 31st year of Ming Emperor Jiajing period (1552). The principal construction Purple Cloud Hall is of the typical wood-framed building in which there are 36 golden pillars and a statue of Jade Emperor was enshrined.

The emperors had always planned and designed the construction of the complex by themselves and sent officials in charge of management of these buildings. The existing constructions are unique in all the Taoist constructions in China in terms of their largest-scale, high level workmanship, compact structure, exquisite decorations, and a great deal of god's statues and sacrifice offering implements. They show the extracts and essence of the ancient Chinese Taoist architecture. Wudang Taoism, Taoist music and the Wudang martial arts have been derived from the Wudang Mountains as part of the traditional Chinese culture.

In 1994, the ancient building complex in the Wudang Mountains was listed as one of the sites of the World Cultural Heritage and the evaluation made by the World Heritage Commission is:

The palaces and temples which form the nucleus of this group of secular and religious buildings exemplify the architectural and artistic achievements of China's Yuan, Ming and Qing dynasties. Situated in the scenic valleys and on the slopes of the Wudang mountains in Hubei Province, the site, which was built as an organized complex during the Ming dynasty (14th–17th centuries), contains Taoist buildings from as early as the 7th century. It represents the highest standards of Chinese art and architecture over a period of nearly 1,000 years.

_1 The Purple Cloud Hall of Wudang Mountains
_2 The Temple of Primal Harmony of Wudang Mountains

4

_3 The acme of Wudang Mountains
_4 The Crown Prince's Slope of Wudang Mountains
_5 The Crown Prince's Slope of Wudang Mountains
_6 The acme of Wudang Mountains

5

6

Lushan National Park

Mount Lushan is situated in the southern part of Jiujiang City of Jiangxi Province, close to the Yangtze River on the north and adjoining the Lake Boyang on the east, covering a total area of 302 km^2. It has more than 90 peaks, the tallest of which is the Great Hanyang Peak, 1,473.4 m above sea level. On the complicated and diverse types of landform, the Mt. Lushan is dotted with a number of valleys, stone caves, waterfalls, brooks and gullies. It has the unique remains of the Quaternary glacier, becoming the original place of the China's Quaternary Glacier Theory.

The landscapes of the mountain are noted for its marvelous, elegant, perilous and magnificent beauties in the world. It has become an age-old mountain well known not only for its beautiful sceneries, but also for its important position as a favorable place to integrate various culture, religion, education and political activities as a whole. The Chinese poetry on landscape originated from it and since ancient times, numerous poets and writers have been inspired to visit the mountain and left over more than 4,000 poems and pieces of descriptive prose to the present time. For example, Li Bai, a poetic genius of the Tang Dynasty, had visited the mountain for five times, and had written 14 poems to laud its beauty. Hui Yuan (334-416), an eminent monk of the Jin Dynasty, had built the Donglin Temple in the mountain, and founded the Sukhavati (Pure Land) Sect of the Buddhism, winning the honor for the mountain in religious circle. The White Deer Cave Academy, built in 940, had been one of the key academic institutes of the Cheng-Zhu Neo-Confucianism. During the 1930s, Mt Lushan became the Summer Capital of the Nanjing KMT government.

At the end of the 19th century, since Edward Selby Little, a British missionary planned and developed the summer resort at Ridge Gulin, the presentation of cultural and scenic establishments in the mountain has shown a tendency of assimilating western learning and cultures. At the beginning of the 20th century, thousands of Christian missionaries from more than 20 western countries swarmed into Mt. Lushan and began building about 1,000 European or American style villas in addition to many churches of different sects, demonstrating a collection of architectural masterpieces of various countries with diversified styles and fashions, including the Gothic Cathedrals, Byzantine style architecture, Japanese style buildings and Islamic mosques. As a result, a huge summer resort with so many villa constructions came into being. Up to the present time, Buddhism, Taoism, Islamism, Christianity and Catholicism all have set up their respective temples or churches in the mountain, adding a brilliant luster to the local landscapes.

As the legend goes, Mt. Lushan had been the "fairyland of immortals." Numerous strange peaks, grotesque rocks, valleys and crags, waterfalls and springs, and stone caves jointly constitute a strikingly beautiful landscape. The outlying boundless rivers and lakes always serve as a cooler to cool down the hot air round the mountains during the hot summer time, providing abundant rainfalls and spreading mists and clouds around them. The mountains are frequently enveloped by mists and clouds with a frequency of up to 191 days a year, creating mysterious and dream-like sceneries.

In 1982, the State Council designated Mt. Lushan as one in the first group of the National Key Scenic Sites. In 1996, UNESCO included it on the list of the World Cultural Landscape. The following evaluation on it was made by the World Heritage Commission:

Mount Lushan, in Jiangxi, is one of the spiritual centres of Chinese civilization. Buddhist and Taoist temples, along with landmarks of Confucianism, where the most eminent masters taught, blend effortlessly into a strikingly beautiful landscape which has inspired countless artists who developed the aesthetic approach to nature found in Chinese culture.

_1 The Embracing Lake Boyang Valley (*Hanbo kou*) at Mount Lushan

1

錦繡谷

_2 The landscapes of Mount Lushan
_3 The waterfall at Mount Lushan

_4 Lake Ruqin at Mount Lushan

Old Town of Lijiang

First built in the late 13th century, the old town of Lijiang is located in the Naxi Autonomous County of Lijiang, Yunnan Province. On the Yunnan-Guizhou Plateau, it is 2,400 m above sea level, stretching over 3.8 km^2. Since ancient times, it has been known for its importance as a commercial, trading and cultural town. The Naxi ethnic minority people account for a majority of the population in the town. About 30 percent of the local residents are still engaged in the making of bronze and silver articles, furs and leather products, and in the textile and wine brewing industries.

The streets in the old town were paved with red breccia along the hillside and close to the streams. During the rainy season, the streets are free from mud or silt, and clean without the nuisance of dust in the dry season. The patterns on the stones are natural and elegant. The Square Street in the downtown is seen as the typical business streets. Over the Yuhe River, a total of 354 bridges of different styles were built, averaging 93 bridges per square kilometre. Among them, the Green-locking Bridge, Big Stone Bridge, Ten-thousands Bridge, South-gate Bridge, Saddle-shaped Bridge, Benevolence and Longevity Bridge are well known ones built during the period of Ming and Qing dynasties (14th –19th centuries A.D.). Having special features most among them is the Big Stone Bridge, about 100 m east of the Square Street. The largest ancient building complex in the town is the Mu's family mansion, formerly the official office and living quarter of the local hereditary headman surnamed Mu. It was first built during the Yuan Dynasty (1271—1368), covering an area of about 30,000 m^2. distributed with 162 large and small houses. In the buildings, 11 horizontal tablets with inscriptions granted by emperors of the past dynasties were decorated over the gates.

With a long history, the old town looks primitive and natural. Its constructions and streets are harmoniously distributed with a unique local style, which has provided significant and valuable heritage for the research of Chinese architectural history. The town is replete with traditional national cultural properties, not only highlighting the prosperity and development of the Naxi minority, but also being important historic remains for the study of human cultural development.

In 1997, the old town of Lijiang was listed as one of the sites of the World Cultural Heritage by UNESCO and the evaluation made by the World Heritage Commission is:

The Old Town of Lijiang, which is perfectly adapted to the uneven topography of this key commercial and strategic site, has retained a historic townscape of high quality and authenticity. Its architecture is noteworthy for the blending of elements from several cultures that have come together over many centuries. Lijiang also possesses an ancient water-supply system of great complexity and ingenuity that still functions effectively today.

2

_1 Looking down at Lijiang
_2 Lanes and streets of Lijiang
_3 Lanes and streets of Lijiang
_4 The Black Dragon Pool at Lijiang

_5 The three-pit well at Lijiang
_6 Buildings at Lijiang
_7 Lanes and streets of Lijiang

桥 咖啡

8

9

_8 Night scene of Lijiang
_9 A dwelling house at Lijiang
_10 Lanes and streets of Lijiang
_11 Night scene of Lijiang

10

Ancient City of Pingyao

The ancient city of Pingyao is located in the Pingyao County of Shanxi Province. It was first built during the period of 827—782 B.C., or the reign of the King Xuan of the Western Zhou Dynasty, some 2,700 years ago. In the third year of the reign (1370) of Emperor Taizu of the Ming Dynasty, a square shaped city was built with its walls some 12 m high, and 6.4 km in circumference. Covering an area of 2.25 km^2, the city has defensive installations such as barbican entrances, crenels, turrets and watchtowers as well as other supporting constructions including the Tower of God's Literature, Terrace of Task-assignment, and Temple of Lord Guan Yu. During the Ming and Qing dynasties, although the city had been repaired and improved, its prototype and former system were essentially preserved intactly. With a population of 42,000 in the city, it is a well- preserved prototype of Chinese ancient city still existing in a region inhabited by Han people.

The city has three treasures: One is its city walls. The walls have 3,000 crenels and 72 turrets to symbolize, it is said, the 3,000 disciples and 72 saints of Confucius. After the vicissitudes through development in about 600 years, it still keeps its former vigorous and robust appearance. The streets, shops and dwelling houses have all maintained their traditional antique layout and styles. The dwelling houses were all built with grey bricks and tiles in the pattern of the courtyard dwelling compound. With a clear-cut axial alignment and symmetrical layout, these houses present a rising and falling outline with the half-hidden and quiet courtyards. The exquisite wood carvings, brick carvings and stone sculpture matched with paper cutting flowers on windows of dwellings also give a lively spectacular. The second treasure is the County-guarding Temple in which the Hall of Ten-thousand Buddhas was built during the Five Dynasties (the 10th century), ranking the third in the type of ancient Chinese wood-framed architecture, dating from more than 1,000 years. The third one is the Double-woods Temple built in the second year of the Emperor Wuping period of the Northern Qi Dynasty (571). More than 2,000 colored statues of figures modeled in a period from Yuan to Ming dynasties (from the 13th century to 17th century) are well preserved in about 10 halls, known as "a depository of color sculpture treasures."

Pingyao City had been one of the birthplaces of the well known "merchants from Jin (Shanxi) Province." In the fourth year of the reign of Emperor Daoguang of the Qing Dynasty (1824), China's first exchange shop, the embryonic form of the modern bank, called "Ri Sheng Chang", was established, whose business had developed into Japan, Singapore, Russia and other countries. During the 100 years of booming development of exchange business, the old city had exerted positive influences on the economic development of modern China.

In 1997, the ancient city of Pingyao was listed as one of the sites of the World Cultural Heritage by UNESCO. The evaluation made by the World Heritage Commission is:

Ping Yao is an exceptionally well-preserved example of a traditional Han Chinese city, founded in the 14th century. Its urban fabric shows the evolution of architectural styles and town planning in Imperial China over five centuries. Of special interest are the imposing buildings associated with banking, for which Ping Yao was the major centre for the whole of China in the 19th and early 20th centuries.

_1 Street scene at Pingyao
_2 Dwelling houses at Pingyao
_3 The ancient city wall of Pingyao

4

5

_4 A dwelling house at Pingyao
_5 An exchange shop at Pingyao
_6 A dwelling house at Pingyao

6

8

_7 Inside view of a dwelling house at Pingyao
_8 Inside view of a dwelling house at Pingyao
_9 Inside view of a dwelling house at Pingyao

9

_10 A screen wall of a dwelling house at Pingyao
_11 An exchange shop at Pingyao

晋元楼

Classical Gardens of Suzhou

"Gardens in South China rank first in the country. But gardens in Suzhou are more beautiful than those in South China." Suzhou is one of the well known historic and cultural cities in China, always noted for its graceful landscape of mountains and waters and its elegant gardens. According to historical documents, the ancient Suzhou city boasted about 200 gardens. Of all the gardens, the Surging Wave Pavilion, the Lion Grove Garden, the Humble Administrator's Garden and the Lingering Garden are the most typical, known as the "Four Noted Gardens," demonstrating the garden-building styles and art of the four dynasties of Song (960-1279), Yuan (1271-1368), Ming (1369-1644) and Qing (1644-1911). Together with the Master-of-Nets Garden, Mountain Villa with Embracing Beauty, Garden of Cultivation, Garden of Couple's Retreat and Garden of Seclusion and Meditation which have also been listed as the sites of the World Cultural Heritage, they are regarded as the outstanding representatives of the classical gardens of Suzhou.

The classic gardens of Suzhou were built not only for living, but also for cultural and artistic entertainments as well as leisure and refreshment, combining living quarters and garden as an enclosed one. People could live, tour and enjoy scenic places in it. The purpose of the garden building was to materialize their life as people in the urban area did and at the same time could enjoy pleasures and mould their temperament in an environment distributed with rustic natural landscapes. The garden owners tried to make use of the gardening and embellishing skills in their construction to fish for fame and compliments. They designed a small garden just for being able to closely and readily enjoy the beauty of sceneries, to create a number of natural sceneries within a limited space, thus creating an aesthetic taste as the saying goes: "hills and woods appear at close hand." The Humble Administrator's Garden, the Lingering Garden, the Master-of-Nets Garden and the Mountain Villa with Embracing Beauty, all embody the Chinese artistic features of workmanship in garden-building. In a limited and small area, the garden builders could employ all kinds of artistic methods to combine some pavilions, terrains, buildings and springs, rocks, flowers and plants as a unit, thus building up an ideal ambience in miniature to produce an effect of "living in a crowded city while breathing fresh air of nature." These gardens reflected the high level of housing civilization in South China at that time.

In 1997, the Humble Administrator's Garden, the Lingering Garden, the Master-of-Nets Garden and the Mountain Villa with Embracing Beauty were listed as the sites of the World Cultural Heritage by UNESCO. In 2000, five more gardens of the Surging Wave Pavilion, Lion Grove Garden, Garden of Cultivation, Garden of Couple's Retreat, and Garden of Seclusion and Meditation as the follow-up, were also listed as the World Cultural Heritage. The evaluation made by the World Heritage Commission is:

Classical Chinese garden design, which seeks to recreate natural landscapes in miniature, is nowhere better illustrated than in the nine gardens in the historic city of Suzhou. They are generally acknowledged to be masterpieces of the genre. Dating from the 11th-19th century, the gardens reflect the profound metaphysical importance of natural beauty in Chinese culture in their meticulous design.

2

_1 The Pavilion of Moonlight and Breeze at the Master-of-Nets Garden
_2 The Fragrant Snow and Azure Cloud Pavilion at the Humble Administrator's Garden
_3 Looking down at the Garden of Seclusion and Meditation

4

5

6

_4 A gate of the Lion Grove Garden
_5 The waterside pavilion at Mountain Villa with Embracing Beauty
_6 An old house at Garden of Couple's Retreat

_7 Crown Peak at the Lingering Garden
_8 Meandering Brook Villa at the Lingering Garden

_9 A waterside corridor and house at the Humble Administrator's Garden
_10 Stone Pavilion at the Mountain Villa with Embracing Beauty
_11 Watery scenery at the Lion Grove Garden
_12 A secluded house amid the Chinese parasol trees and bamboos at the Humble Administrator's Garden
_13 The Hall of Cool Breeze Pond at the Lingering Garden
_14 Courtyard of the Garden of Joy

10

11

9

12

13

東安中峯
眘谷題名

_15 Water scenery alongside the veranda of the Garden of Seclusion and Meditation
_16 Pavilion of Stay-to-Listen at the Humble Administrator's Garden
_17 Inside view of Jade Pot and Ice Chamber of the Humble Administrator's Garden
_18 Veranda alongside the lake at the Humble Administrator's Garden

玉壶冰

Summer Palace in Beijing

The Summer Palace is located in the northwestern part of the Beijing City. Originally it had been an auxiliary palace for emperors, first built in 1153, or the first year of the Emperor Zhenyuan period of the Jin Dynasty. During the Ming Dynasty, it was renamed as Garden of Excellent Hills. Then it was expanded in 1750 or the 15th year of the reign of the Qing Emperor Qianlong, and was again renamed as Garden of Clear Ripples when the renovation was completed after 15 years. In 1860 when the British-French Allied Forces invaded into Beijing, the garden was destroyed on fire. From 1886 to 1891, the Empress Dowager Cixi, reconstructed the garden with huge amount of money embezzled from the navy fund. And it was once again renamed as the Summer Palace after the construction was completed, meaning "taking care of oneself in peace." In 1900 when the Eight-Power Allied Forces attacked Beijing, the garden once more suffered the scourge of war. In 1902, the Empress Dowager again allotted funds to repair the damaged garden.

On the basis of the garden design

concepts of divine right of emperors and the sovereign descending the world, the Summer Palace was designed as an imperial garden ingeniously making use of the surrounding natural sceneries. It boasts more than 3,000 constructions such as palaces, halls, temples and gardens, distributed in three sections: administrative area, living quarters, leisure and touring areas. Taking the Kunming Lake and the Pavilion of the Buddhist Incense on the Longevity Hill as its key scenic spots, against a backdrop of the Back Hill and the Back Lake, the garden presents a picture of green hills and blue waters, matched with long and winding corridors, and some other graceful pavilions and gorgeous halls. Its main style of art embodies the motif of "creating a humanized nature" on the basis of the garden's natural landscape. It contains either the grace of natural wilderness or the awe-inspiring supreme authority of emperors; it diffuses a harmonious religious and secular atmosphere, and develops an agreeable ambience either for a leisure and refreshment life or for dealing with government affairs. Visitors could communicate their feelings with the graceful landscape and immerse themselves in the admiration for the beauties. The Summer Palace enjoys a top honor in the world history of the gardening art.

In 1961, the State Council designated the Summer Palace as one in the first group of the key cultural relics under state protection. In 1998, the palace was listed as the World Cultural Heritage by UNESCO and an evaluation made by the World Heritage Commission is:

The Summer Palace in Beijing – first built in 1750, largely destroyed in the war of 1860 and restored on its original foundations in 1886 – is a masterpiece of Chinese landscape garden design. The natural landscape of hills and open water is combined with artificial features such as pavilions, halls, palaces, temples and bridges to form a harmonious ensemble of outstanding aesthetic value.

_1 The Kunming Lake and the Longevity Hill at the Summer Palace
_2 The Marble Boat at the Summer Palace
_3 West Causeway Bridge at the Summer Palace
_4 A distant view of the Hill of Jade Spring seen from the Summer Palace

4

6

_5 The Pavilion of the Buddhist Incense at the Summer Palace
_6 The Pavilion of the Buddhist Incense at the Summer Palace
_7 Looking down at the Kunming Lake
_8 The Back Lake at the Summer Palace

7

8

9

_9 Revolving Scripture Repository at the Summer Palace
_10 The Back Lake at the Summer Palace
_11 Boundary for Pilgrims at the Summer Palace

10

衆香界

_12 The Kunming Lake at the Summer Palace

Temple of Heaven in Beijing

Situated at the southern end of the imperial city of Beijing, the Temple of Heaven was a sacred place for the emperors of Ming and Qing dynasties to worship Heaven and pray for good harvest. In ancient China, only the emperor, the son of Heaven as he claimed himself, had the exclusive privilege to worship Heaven. The Temple of Heaven in Beijing was first built in 1420 A.D., or the 18th year of the Yongle period of the Ming Dynasty. At that time, all the ritual ceremonies of worshipping Heaven and Earth took place there, called "Temple of Heaven and Earth." In 1530, Emperor Jiajing of Ming Dynasty established another temple in the northern suburbs of Beijing, called Square Pool Temple, to worship Earth only. Since that time, this place had been in service for worshipping Heaven.

The Temple of Heaven in Beijing is the largest construction complex in the world for worshipping Heaven. It is also an imperial garden used for holding such worshipping ceremonies, largest on scale and highest in rank in China. With its exquisite design, ingenuity of construction, and unique style, it has been hardly seen before in the world. Extending over 2.73 million m^2, it is a bit smaller than that of the Summer Palace. The whole layout was in the shape like the Chinese character "回", divided into two parts: The inside temple and outside temple enclosed with respective walls. Its design including the general layout and detail embellishments emphasized the motif: "the will of god."

The two parts of the Temple of Heaven are Circular Mound Altar and the Hall of Prayer for Good Harvests, distributed at the south and north ends of the central axial alignment linked by a Red Stairway Bridge which is 360 m long, 28 m wide and some 2 m high. The two great halls were designed in a circular shape on a plane surface, symbolizing the traditional Chinese understanding of the universe, that is, "Heaven is a dome and earth square." The Circular Mound Altar was specially built for emperors to worship Heaven. It is a multitier high terrace.Its balustrades and watching pillars were built in odd numbers (or *yang* numbers, or heavenly numbers, namely the number nine, and the multiple of nine). The surface of the altar was paved with nine tiers of stone boards. Except the circular shape of board in the center, the boards paved on each circle around the center were shaped like a fan. The board number of each circle was also designed to be nine or multiple of nine, identical to the nine storeys of Heaven as the legend says. Various sacrifice-offering ceremonies taking place in the Temple of Heaven were not only just a ritual ceremony in form of paying homage to Heaven, but also an occasion for emperors to show off their utmost authorities and paramountcy.

In 1961, the State Council announced that the Temple of Heaven was listed as one of the key cultural relics under state protection. In 1998, UNESCO included it on the list of the World Cultural Heritage. The evaluation made by the World Heritage Commission is:

The Temple of Heaven, founded in the first half of the 15th century, is a dignified complex of fine cult buildings set in gardens and surrounded by historic pine woods. In its overall layout and that of its individual buildings, it symbolizes the relationship between earth and heaven – the human world and God's world – which stands at the heart of Chinese cosmogony, and also the special role played by the emperors within that relationship.

2

1

3

_1 Hall of Prayer for Good Harvests at the Temple of Heaven
_2 Stone archway at the Temple of Heaven
_3 Imperial Vault of Heaven at the Temple of Heaven
_4 Hall of Prayer for Good Harvests at the Temple of Heaven
_5 Inside view of the Hall of Prayer for Good Harvests

4

5

_6 Imperial Vault of Heaven at the Temple of Heaven
_7 Lattice Star Gates of the Temple of Heaven
_8 Inside view of the Imperial Vault of Heaven at the Temple of Heaven

皇天上帝

_9 Inside view of the Imperial Hall of Heaven at the Temple of Heaven
_10 The caisson of the Hall of Prayer for Good Harvests at the Temple of Heaven
_11 The scene of sunrise at the Temple of Heaven

Dazu Rock Carvings

The Dazu Rock Carvings were the religious cliffside rock carvings produced during the period of the Tang and Song dynasties, mainly describing the religious legends and stories. And the statues on the North Mountain and the Mountain of Treasure Top (Baoding) are the best known.

Dazu cliff statue-carvings are centralized in the area of Dazu County of Chongqing Municipality, first cut during the Emperor Yonghui period of the early Tang Dynasty (649), and developed throughout the later Tang and the Five Dynasties (907-959) to its prime stage in the Northern and Southern Song dynasties. During the period of Ming and Qing dynasties (14th –19th centuries), the number of carvings increased and it eventually formed a huge complex of the best artistic carving products which are regarded as the masterpiece of the later stage of the grotto art of China, enjoying an honor in parallel with those of Yungang Grottoes, Longmen Grottoes and Mogao Caves.

Dazu rock carvings involved more than 100,000 stone statues distributed in some 70 locations, mainly on the North Mountain, Mountain of Treasure Top, Nanshan Mountain, Shizhuan Mountain, and the Stone Gate Mountain.

The rock carvings on the North Mountain are located 2 km northwest of the Dazu County, first produced in the first year of Tang Jingfu period, and lasted until the Shaoxing period (1162) of the Southern Song Dynasty, making nearly ten thousand stone carving statues on cliffs. The work was mainly supported by donations from laymen. The statues were modeled to depict life stories about those eminent Buddhas who were very popular at the time, as a means to promote the secular education of Buddhism. With their exquisite craftsmanship, high level of skills, and elegant modeling, the statues on the North Mountain demonstrate the history of folk beliefs in Buddhism and the changes of artistic flavors of stone carvings in the period from the 8th century to the 14th century.

The stone carvings on the Mountain of Treasure Top were distributed in an area 15 km northeast of the Dazu County. They were first cut in the 6th year of the Chunxi period of the Southern Song Dynasty (1179) and lasted till the 9th year of the Chunyou period of the same dynasty (1249). These carving statues centered on the Temple of Imperial Longevity including 13 groups of statues scattered in an area involving the Giant Buddha Bend and Little Buddha Bend with a total of about ten thousand statues, focusing on the life stories of eminent monks of the Esoteric Sect of Buddhism. The amalgamation of these statues looks like a large Buddhist holy place, displaying the carving masterpieces of the Song Dynasty (960-1278).

The Dazu rock carvings are noted for its gigantic work, elegant craftsmanship, wide coverage of themes, richness in intentions and being preserved in good condition. As the rock carving masterpieces bearing the characteristics of the Buddhism, Taoism and Confucianism, they have become a bright pearl of the Chinese grotto artistic products, with an irreplaceable value in the research of history, art and science.

In December 1999, UNESCO listed the Dazu rock carvings as the World Cultural Heritage and the evaluation made by the World Heritage Commission is:

The steep hillsides of the Dazu area contain an exceptional series of rock carvings dating from the 9th to the 13th century. They are remarkable for their aesthetic quality, their rich diversity of subject matter, both secular and religious, and the light that they shed on everyday life in China during this period. They provide outstanding evidence of the harmonious synthesis of Buddhism, Taoism and Confucianism.

2

_1 Dazu stone statues of the holy process of Sakya being blown out
_2 Stone statues at Dazu
_3 Stone statues at Dazu
_4 Stone statue of Avalokitesvara at Dazu
_5 Stone statues at Dazu

4

瑜伽部主總持王
佛日光輝
常轉

_6 Stone statues at Dazu
_7 Stone statues at Dazu
_8 Stone statues at Dazu
_9 Stone statues at Dazu
_10 Dazu stone statues of the holy process of Sakya being blown out

Mount Qingcheng and the Dujiangyan Irrigation System

Situated at the mountain side 15 km southwest of Dujiangyan City, Sichuan Province, Mount Qingcheng has long since enjoyed the honor of "a beautiful and secluded Qingcheng on earth," for its exuberant evergreen vegetation. Since it was the important birthplace of the Chinese Taoism and the mother mountain of the Taoist Master Sect, the Mount Qingcheng has become a sacred mountain for its beautiful natural landscape and a great number of constructions in Taoist architectural style. In the mountain, the 10 or more Taoist temples and monasteries, centering on the Cave of Taoist Master, include the Fortune Founding Hall, the Hall of Superior Heaven, Hall of Patriarch, Hall of Perfect Splendor, Pavilion of Immortality, Hall of Jade Purity, and the Sun-facing Cave. The Fortune Founding Hall was first built in the 18th year of the Kaiyuan period of Tang Dynasty (730 A.D.), and in the three rows of the hall buildings, the eminent Taoist priests and many deities were enshrined respectively. The antithetical couplet scrolls written in 394 Chinese characters and decorated on the pillars of the hall are praised by visitors as "a marvel of Qingcheng." Its constructions, fully embodying the Taoist idea of pursuing for a natural being, were ingeniously built along the lines of the landform, and the embellishments were designed in accordance with the Taoist ideas of seeking for auspiciousness, longevity, and immortality.

Dujiangyan was an ancient irrigation construction and has long been used and lasted till today, located in the upper reaches of the Minjiang River west of Dujiangyan City in Sichuan Province. It was first built during the period from 256 B.C. to 251 B.C. under the direction of Li Bing, a satrap of the Sichuan (Shu) Commandary of Qin State during the Warring States period and his son. With the renovations and repairs throughout the passed ages of more than 2,000 years, the irrigation system has since played a great role. The irrigation system can be divided into two major sub-systems: the main construction includes the fish mouth (water distribution), sand-discharging weir (overflowing flood and draining sand), and bottle mouth (channeling water), in addition to other supporting constructions such as the exterior and interior Jingang Dyke, and Herringbone Dyke. The whole system has been used mainly for introducing water for irrigation, preventing from floods, draining sands and silts, and for water navigation or water supply to cities. The Chengdu Plain irrigated by the system is well known as "the land of plenty."

In 2000, Mount Qingcheng and the Dujiangyan irrigation system were listed as one of the sites of the World Cultural Heritage by UNESCO and the evaluation made by the World Heritage Commission is:

Construction of the Dujiangyan irrigation system began in the 3rd century B.C. This system still controls the waters of the Minjiang River and distributes it to the fertile farmland of the Chengdu plains. Mount Qingcheng was the birthplace of Taoism, which is celebrated in a series of ancient temples.

2

_1 Temple gate of Mount Qingcheng
_2 Hall of Superior Heaven at Mount Qingcheng
_3 Hall of Patriarch at Mount Qingcheng
_4 A scenic pavilion at Mount Qingcheng

_5 Temple of Two Nobilities at Dujiangyan
_6 Hall of Superior Heaven at Mount Qingcheng
_7 Waterfall at Mount Qingcheng
_8 A full view of Dujiangyan

Longmen Grottoes

Situated in the Longmen Gorges between the east and west cliffs at the Yihe River, southern suburbs of Loyang City in Henan Province, the Longmen Grottoes extends over 1 km from south to north. It has a long history of more than 1,500 years. First cut during the period of the Emperor Xiaowen of Northern Wei Dynasty (471-477), the building of the grottoes had lasted for more than 400 years until the Northern Song Dynasty. With its existing grottoes totaling about 1,300 and 2,345 cave shrines, 3,600 stone-carved tablets with inscriptions, some 50 pagodas, and 97,000 statues of Buddha, it boasts the largest quantity of grottoes in China. The most typical caves include the Ancient Sunny Cave, Binyang Cave, Lotus Cave, Medical Prescription Cave built during the Northern Wei period, and the Latent Stream Cave, Scripture

Reading Temple, Ten-thousand Buddha Cave, Fengxian Temple and the Dawan Five-blessing Cave, built during the Tang period.

The Ancient Sunny Cave was cut in the earliest time with richest intentions. Most of its enshrined statues of Buddha were inscribed with comments and notes, recording the names of the statue-makers, the date of the work and the cause of production, providing valuable materials for the study of the calligraphic and carving art of the Northern Wei Dynasty. The "Twenty Articles of the Longmen Cave," cream of the Wei tablet style, have been regarded as an important landmark in the history of the Chinese calligraphic art.

The Binyang Cave was cut during the Northern Wei Dynasty (386--512), throughout a period of 24 years. There stand 11 statues of Buddha, and the principal one is the statue of Sakyamuni. With its natural complexion and delicate features, this statue has won the fame of a cream of the stone-carved products during the middle of the Northern Wei Dynasty. In front of the principal statue, a pair of robust stone lions stands. And on both sides of the statue stand two disciples and two Bodhisattvas, smiling and staring, and wearing an honest, sincere and kind expression.

The Fengxian Temple is the largest cave of all, a typical one built with the style of the Tang Dynasty (618-904). The temple is 30 m long and 30 m wide, in which stands a giant statue of Buddha Vairocana, 17.14 m tall, looking lively, elegant and plump. It is said Vairocana was the nirmanakaya of Wu Zetian, the only Chinese empress in ancient times.

The huge and magnificent Longmen Grottoes with their exquisite and delicate stone-carved statues in caves have been honored as one of the greatest treasure houses of classic art in the world. In 2000, they were listed by UNESCO as one site of the World Cultural Heritage and the evaluation made by the World Heritage Commission is:

The grottoes and niches of Longmen contain the largest and most impressive collection of Chinese art of the late Northern Wei and Tang Dynasties (316-907). These works, entirely devoted to the Buddhist religion, represent the high point of Chinese stone carving.

_1 The full view of the Fengxian Temple at the Longmen Grottoes
_2 The Buddha Vairocana at the Longmen Grottoes
_3 Statues at the Longmen Grottoes

_4 The image of Sakyamuni in Binyang Cave
_5 Statues at the Longmen Grottoes
_6 Statues at the Longmen Grottoes

Imperial Tombs of the Ming and Qing Dynasties

The assemblage of imperial tombs of the Ming, the Later Jin and the Qing dynasties are called the imperial tombs of the Ming and Qing dynasties. The period of Ming and Qing (1368-1911) saw the thriving stage of the imperial tomb building history in China. The founding emperor Zhu Yuanzhang of the Ming Dynasty revised the past tomb system by setting up more sacrifice offering and ritual installations, more buildings and courtyards with canopy-style roofs and eliminating the temple palace. The Qing Dynasty inherited the Ming system, but paid more attention to the geomancy and ambience of the mausoleum, wishing it to be built by creating an environment in which the tomb and the surrounding mountains and waters would combine into unbroken harmony as "a unity of Heaven and man." And it also stressed the importance of the seniority and superiority of the demised in the family and instituted the order system for emperors, empresses, and imperial concubines and

their tombs would be built in a more beautiful and imposing manner.

The imperial tombs had been planned to be established through a unified institution and form in one enclosed area with a single entrance. Starting from the Red Gate, people can enter it through the stone sculptures standing on the both sides of the sacred passage, stone tablet pavilions, ornamental pillars, and then come to the respective section of a certain tomb. Its general layout was in order as follows: The five arched stone bridge, a memorial archway, a tablet pavilion, the three arched stone bridge, a great platform, the gate of palace, the Hall of Grand Grace with its right and left wings, and then come to the stone beam bridge, a platform, a glazed gate, the five offerings, a Square City (with a bright tower erected), the City of Crescent Moon, the City of Treasures, and the canopy of the tomb. A rigid order system had been instituted in the mausoleum for emperors, empresses, princes, princesses, and imperial concubines.

The assemblages of tombs which have been listed as one site of the World Cultural Heritage include the Eastern Royal Tombs of the Qing Dynasty in Hebei Province (involving the five emperors' tombs, being listed in 2000); the Western Mausoleums of the Qing Dynasty in Hebei Province (containing four emperors' tombs, listed in 2000); the Ming Xianling Mausoleums in Hubei Province (tombs of the parents of Ming Emperor Shizong, listed in 2000); the Ming Tombs in Beijing (listed in 2003); the Ming Xiaoling Mausoleums in Jiangsu Province (tombs of Emperor Zhu Yuanzhang, founder of the Ming Dynasty, and his empress and concubines (listed in 2003); the three mausoleums in Liaoning Province (including Fuling Tomb of Emperor Nurhaci and his empress, Zhaoling Tomb of Abahai and his empress, and Yongling Tomb of ancestors of Nurhaci, listed in 2004).

In 2000, UNESCO included the imperial mausoleums of Ming and Qing on the list of the World Cultural Heritage

2

and in 2003 and 2004, more imperial tombs listed above were also included. The evaluation made by the World Heritage Commission is:

It represents the addition of three Imperial Tombs of the Qing Dynasty in Liaoning to the Ming tombs inscribed in 2000 and 2003. The Three Imperial Tombs of the Qing Dynasty in Liaoning Province include the Yongling Tomb, the Fuling Tomb, and the Zhaoling Tomb, all built during the 17th century. Constructed for the founding emperors of the Qing Dynasty and their ancestors, the tombs follow the precepts of traditional Chinese geomancy and fengshui theory. They feature rich decoration of stone statues and carvings and tiles with dragon motifs, illustrating the development of the funerary architecture of the Qing Dynasty. The three tomb complexes, and their numerous edifices, combine traditions inherited from previous dynasties and new features of Manchu civilization.

3

_1 The sacred passage of the Xiaoling Tomb at the Eastern Royal Tombs of the Qing Dynasty
_2 The memorial archway of the Dingling Tomb at the Eastern Royal Tombs of the Qing Dynasty
_3 The Great Hall of the Changling Tomb at the Ming Tombs in Beijing
_4 An ornamental pillar at the Ming Tombs in Beijing
_5 A distant view of Chongling Tomb at the Western Mausoleums of the Qing Dynasty

5

6

7

8

_6 A distant view of Jingling Tomb at the Eastern Royal Tombs of the Qing Dynasty
_7 A stone tablet pavilion at the Eastern Royal Tombs of the Qing Dynasty
_8 A stone warrior at the Ming Tombs in Beijing
_9 The great stone archway at the Eastern Royal Tombs of the Qing Dynasty
_10 A distant view of Jingling Tomb at the Eastern Royal Tombs of the Qing Dynasty

12

_11 A decorative figurine on the gate of the underground palace of the Yuling Tomb of the Eastern Royal Tombs of the Qing Dynasty
_12 Looking down at the Changling Tomb of the Western Mausoleums of the Qing Dynasty
_13 Intense Memorial Gate at the Zhaoling Tomb of the Qing Dynasty in Shenyang
_14 The base of the Great Hall at the Fuling Tomb of the Qing Dynasty in Shenyang
_15 A tablet pavilion at the Fuling Tomb of the Qing Dynasty in Shenyang
_16 Turret at the Fuling Tomb of the Qing Dynasty in Shenyang
_17 Temple Gate of the Zhaoling Tomb of the Qing Dynasty in Shenyang

14

15

13

16

17

18

_18 Compound of Intense Memorial Hall at the Zhaoling Tomb of the Qing Dynasty in Shenyang
_19 The base of the Intense Memorial Hall at the Fuling Tomb of the Qing Dynasty in Shenyang
_20 The memorial archway at the Zhaoling Tomb of the Qing Dynasty in Shenyang

Ancient Villages in Southern Anhui—Xidi and Hongcun

Xidi and Hongcun villages, situated at Yixian County in the southern part of Anhui Province, are the most two typical villages of southern Anhui dwellings. They are well known in the world for the beautiful rural sceneries in a sequestered land of peace, well-preserved formation of villages, dwellings with exquisite craftsmanship in structure and the vestiges of their abundant, colorful historical and cultural traditions.

The Xidi Village was first built in the Huangyou period of the Northern Song Dynasty (1049-1054). The boat-shaped village has 122 antique dwellings still kept intact. Today, more than 1,000 inhabitants of 300 households are living in the village. Surrounded by mountains and two streams running through it, all the streets and lanes in the village were paved with greenish slabs. Its houses were mostly constructed with wood frames together with bricks and decorated with wood and brick engravings, and the layout was well matched with the streets and roads. The simply and lightly painted houses were well spaced with each other, and are regarded as the typical examples of the Anhui-style constructions.

The Hongcun Village was first built in the first year of the Shaoxing period of the Southern Song Dynasty (1131). Stretching over 190,000 km^2, it has 137 antique houses built during the period of Ming and Qing (1368-1911). Being built on a high mountainous area, the houses are often enveloped by mists and clouds, presenting to view as if they were drawn with thick and dark ink or with a freehand brushwork in a traditional Chinese painting, so, it is crowned with the honor of "a village in a traditional Chinese painting." All the houses are built with whitewash walls and dark tiles, neatly and orderly distributed. Chengzhi Hall, the most sumptuous of them, is praised as an "imperial palace of the folks."

The site, layout and form of buildings of the two villages were all based on the geomancy described in the *Book of Changes*, thus making the elegant dwellings have a close link with nature, and fully embodying the traditional philosophical ideology of "the unity of Heaven and man." The unique water system in the two villages is an exemplary one of the water conservancy facilities, which combines practical utility with an aesthetic taste. Particularly, the "cow-shape" water network in Hongcun Village was a distinguished invention, ingeniously making use of local natural conditions to improve the environment of settlement.

In 2000, these two villages were listed as one of the sites of the World Cultural Heritage by UNESCO and the evaluation made by the WHC is as follows:

The two traditional villages of Xidi and Hongcun preserve to a remarkable extent the appearance of non-urban settlements of a type that largely disappeared or was transformed during the last century. Their street plan, their architecture and decoration, and the integration of houses with comprehensive water systems are unique surviving examples.

_1 Archway for Department Magistrate Hu Wenguang at Xidi Village

_2 Lunar Palus of Hongcun Village

1

2

3

_3 The Huizhou ancestral hall

_4 Step gable of a dwelling house at Hongcun Village
_5 Street view of Hongcun Village

_6 The Huizhou Hu Family's ancestral hall

_7 Entrance of the Xidi Village
_8 Memorial Archway of Tangyue
_9 The Chengzhi Hall of Hongcun Village

7

8

9

_10 A distant view of Hongcun Village

Yungang Grottoes

Yungang Grottoes are situated on the north cliff of Mt. Wuzhou, 16 km west of the Datong City, Shanxi Province. First cut in the second year of the Xing'an period of the Northern Wei Dynasty (453), most of the grottoes had been finished before the capital of Northern Wei was moved to Luoyang (494), and the statue-carving project lasted until the period of Zhengguang (520-525). The grottoes were cut along the sides of mountain, extending about 1 km long. Noted for its grandeur on scale and rich in contents, the grottoes were the first of this kind cut under the direct management of the imperial court and nobles.

The grottoes were categorized into three groups. The greatest statue of Buddha stands 17 m high while the smallest only several centimetres tall. The largest cave is the Cave No. 6, measuring 20 m high from the ground to its ceiling. The east group of caves involves mostly pagodas, also called pagoda caves. Of the central group of caves, the Five Caves Covered by Clouds were cut in the earliest time and are the most imposing ones with a simple and vigorous flavor of the Western Regions. The west group of caves involves mostly products modeled after the capital was moved to Luoyang. In spite of the small size of caves, the statues were delicately and beautifully cut and in moderately proportions, so they have been set as the best models of the grotto art in the northern part of China. They were also regarded as the origin of the type of "bony and slender statues," a school of the modeling art. The statues of Buddha, guardians, and the Flying Apsaras were all vividly modeled. And the patterns on the columns of pagodas were exquisitely and delicately engraved. All these artistic products were made by inheriting the styles of the art creams of the Qin and Han realistic products, at the same time set examples of romantic art style for the following Sui and Tang dynasties. Those stone carving of acrobatic, dancing and other performances left over reflect the popularity of the Buddhist ideology in then social life.

The grottoes have figuratively recorded the evolution of the Chinese Buddhist art developed from that of India and the Middle Asia. Moreover, the statue modeling changed gradually from foreign styles into Chinese styles, becoming an art of secularization and nationalization. Various styles of Buddhist image-modeling were merged and mixed up, thus creating the "Pattern of Yungang" as a turning point in the Buddhist art development in China. The layout and decorations in the caves during the later stage of its development specially displayed the strong flavors and remarkable styles of the Chinese architectural elements and showed the further deepening of the Buddhist art with the Chinese national characteristics.

In 2001, The Yungang Grottoes were listed as the World Cultural Heritage by UNESCO and the evaluation made by the World Heritage Commission is as follows:

The Yungang Grottoes, in Datong city, Shanxi Province, with their 252 caves and 51,000 statues, represent the outstanding achievement of Buddhist cave art in China in the 5th and 6th centuries. The Five Caves created by Tan Yao, with their strict unity of layout and design, constitute a classical masterpiece of the first peak of Chinese Buddhist art.

_1 Statues of Buddha at Yungang Grottoes
_2 Statues of Buddha at Yungang Grottoes
_3 Outside view of Yungang Grottoes

1

2

3

4

5

6

7

_4 Giant Buddha statue of Yungang Grottoes
_5 Statues of Buddha at Yungang Grottoes
_6 Statues of Buddha at Yungang Grottoes
_7 Statues of Buddha at Yungang Grottoes

_8 Statues of Buddha at Yungang Grottoes

Capital Cities and Tombs of the Ancient Koguryo Kingdom

The capital cities and tombs of the ancient Koguryo Kingdom were the remains of the earlier stage of the Koguryo Kingdom, distributed around Ji'an City of Jilin Province.

The Koguryo had been a state power of an ethnic minority group in the frontier area of ancient China. In the second year of the ruling period of Emperor Yuandi of the Western Han Dynasty (37 B.C.), Zhu Meng, a native of Fuyu, founded the state in the Koguryo County of Xuantu Commandery (in the present time Xinbin County of Liaoning Province). Later, it established its capital in Qishenggu City (now Wunu Mountain City of Huanren County of Liaoning Province). In the third year of the Emperor Pingdi period of Western Han (3 A.D.), it moved the capital to Guonei City and built the Weinayan City (both in today's Ji'an County of Jilin Province), then it again moved its capital to Pyongyang in the fourth year of the Shiguang period of Northern Wei Dynasty (427 A.D.). In the year 688, the Koguryo Kingdom was thrown down by the allied forces of the Tang Dynasty and the Silla Kingdom in the Korean Peninsula.

As one of the local states of ethnic minority with its own characteristics and influences, the Koguryo Kingdom had had a glorious history of its own, playing an important role in the history of the Northeastern Asia. Most of its historical relics have been discovered in Jilin and Liaoning provinces in China, providing irreplaceable evidence to an episode of its history, and possessing important historical and cultural values. The capital cities and tombs of ancient Koguryo Kingdom were listed in the World Heritage involving the Wunu Mountain City, Guonei City, Wandu Mountain City, 14 tombs of kings, 26 tombs of nobles, the Tablet of King Haotai and the No. 1 tomb accompanying to the General Tomb .

In July 2004, at the Suzhou session of the World Heritage Commission of UNESCO, these cultural relics of the kingdom in China together with its group of tombs discovered in Pyongyang and Nampo in DPRK were listed respectively as the World Cultural Heritage. The evaluation made by the WHC is:

The site includes archaeological remains of three cities and 40 tombs: Wunu Mountain City, Guonei City and Wandu Mountain City, 14 tombs are imperial, 26 of nobles. All belong to the Koguryo culture, named after the dynasty that ruled over parts of northern China and the northern half of the Korean Peninsula from 277 BC to 668 AD. Wunu Mountain City is only partly excavated. Guonei City, within the modern city of Ji'an, played the role of a 'supporting capital' after the main Koguryo capital moved to Pyongyang. Wandu Mountain City, one of the capitals of the Koguryo Kingdom, contains many vestiges including a large palace and 37 tombs. Some of the tombs show great ingenuity in their elaborate ceilings, designed to roof wide spaces without columns and carry the heavy load of a stone or earth tumulus (mound), which was placed above them.

2

_1 The General Tomb
_2 A watch tower of the royal city of Koguryo
_3 Tombs of the nobles of Koguryo

4

_4 Colorful paintings for tombs of Koguryo
_5 Colorful paintings for tombs of Koguryo
_6 Tombs of the nobles of Koguryo

5

6

Historic Centre of Macao

The historic center of Macao is composed of 22 buildings and nearby eight squares in the Macao Peninsula, taking the downtown of the old city as its center. It encompasses an area starting from the Colina da Guia in the east to the Rua do Quimaraes close to the Inner Harbor in the west and from the Colina da Barra in the south to the Luis de Camoes Garden in the north.

The site of the historic center of Macao in a bid for World Cultural Heritage includes the A-Ma Temple, Moorish Barracks, Mandarin's House, St. Lawrence Church, St. Joseph Seminary and Church, Dom Pedro V. Theatre, Sir Robert HoTung Library, St. Augustine Church, Leal Sedano Building, Sam Kai Vui Kun Temple (Kuan Tai Temple), Holy House of Mercy, Cathedral, Lou Kau Mansion, St. Dominic Church, Ruins of St. Paul Catholic Church, Na Tcha Temple, the Old City Walls, Mount Fortress, St. Anthony Church, Casa Garden, Old Protestant Cemetery, Guia Fortress (including Guia Chapel and Guia Lighthouse), and seven other squares close to these historic buildings including Barra Square, Lilau Square, Dom Pedro V. Square, Senado Square, St. Dominic Square, Jesuit Square, and the Luis de Camoes Square.

The historic city center contains the oldest, largest and best preserved building complex in China, in both the western and oriental styles. They include the remains of China's oldest church and convent, the oldest protestant cemetery, the oldest construction complex of western fortresses, the first western-style theatre, the first modern lighthouse and the first western-mode college. As the first concession (least territory) in an eastern asian country to be supposedly least to a European country, it bears witness to the history of the exchange and co-existence of Chinese and western cultures in Macao over the past 400 years. Most of the constructions bear the features of blending Chinese architectural elements with the western ones and they still maintain their former functions in good conditions.

2

3

_1 Santa Casa da Misericordia and Senado Square in Macao
_2 Senado Square in Macao
_3 Monte Fortress in Macao
_4 Antique architecture

At the 29th session of the World Heritage Commission of UNESCO, held on July 15 in 2005, the historic center of Macao was unanimously approved by the 21 member states of the commission to be listed as the World Cultural Heritage. The evaluation made by the commission is:

Macao, a lucrative port of strategic importance in the development of international trade, was under Portuguese administration from the mid-16th century until 1999, when it came under Chinese sovereignty. With its historic street, residential, religious and public Portuguese and Chinese buildings, the historic centre of Macao provides a unique testimony to the meeting of aesthetic, cultural, architectural and technological influences from East and West. The site also contains a fortress and a lighthouse, the oldest in China. It bears witness to one of the earliest and longest-lasting encounters between China and the West, based on the vibrancy of international trade.

5

6

_5 Santa Casa da Misericordia
_6 Antique architecture in Macao
_7 Ruins of St. Paul Catholic Church

_8 Outside view of the St. Lawrence Church
_9 Guia Lighthouse in Macao
_10 Inside view of the St. Lawrence Church

Yin Xu

Yin Xu or Yin Ruins, situated in the western part of Anyang City of Henan Province, across the south and north banks of the Huanhe River, called "Bei Meng" in ancient times, is the remains of the capital of the later period of Shang Dynasty (1250 B.C.—1050 B.C.). Covering an area of about 36 km^2, it was the first remains of a capital in Chinese history, having been testified by historical documents and inscriptions on bones as well as some other excavated historical relics. It has witnessed a long history of 3,300 years until now.

In the oracle bone inscriptions, Yin Xu was also known as a "big trading county." At the beginning of the 20th century, Yin Xu happened to be discovered only because its buried animal bones had been stolen by some unknown diggers. In 1928, archeologists began to excavate officially on the site. The huge and densely contained remains include relics of palaces, temples, and a hand-crafting workshop area for copper casting, bone-tool making and pottery making, and living quarters, the imperial tombs area, and a graveyard of common people. The unearthed relics also include enormous bronze ware, jade articles, horn implements and pottery. Some fine and elegant bronze ware for ritual ceremonies such as the Simuwu Ding are also included. Also unearthed are 150,000 pieces of inscribed bones with oracles including some 5,000 single words that are the earliest written characters of China. All these abundant cultural relics of the Yin ruins reflected the highly developed bronze civilization of the Shang Dynasty.

Yin Xu is the site where the oracle bone inscriptions are first discovered, which provides important and rich materials for the study of the history of late Shang Dynasty. Of the world's four key ancient writing systems, the age-old Chinese character system represented by the oracle bone inscriptions, is the only one that has been passed down to the present time through numerous changes in the past thousands of years, having written the extensive and profound Chinese history of civilization. In the Yin ruins, about 110 relics of palaces and temples have been discovered, some of which were laid out in groups, such as the ancestor's temples or the community's altars for sacrifice ceremonies, being regarded as the embryonic form of

4

5

_1 Entrance of the Palace and Royal Ancestral Shrines Area of Yin Ruins
_2 Oracle bones
_3 The chariot and horse pit
_4 Remains of the humans buried with the demised
_5 The pit of funerary objects
_6 The pit of oracle bones
_7 The chariot and horse pit

3

the later regularized architectural system and pattern for Chinese imperial palace constructions in the following dynasties.

In March 1961, the State Council designated the Yin Xu site in the first group of the key cultural relics under state protection. On July 13, 2006, the Yin Xu was included on the list of World Cultural Heritage for its universal distinguished value and good management and exhibition. The evaluation made by the WHC is as follows:

The archaeological site of Yin Xu, close to Anyang City, some 500 km south of Beijing, is an ancient capital city of the late Shang Dynasty (1300 - 1046 BC). It testifies to the golden age of early Chinese culture, crafts and sciences, a time of great prosperity of the Chinese Bronze Age. A number of royal tombs and palaces, prototypes of later Chinese architecture, have been unearthed on the site, including the Palace and Royal Ancestral Shrines Area, with more than 80 house foundations, and the only tomb of a member of the royal family of the Shang Dynasty to have remained intact, the Tomb of Fu Hao. The large number and superb craftsmanship of the burial accessories found there bear testimony to the advanced level of Shang crafts industry. Inscriptions on oracle bones found in Yin Xu bear invaluable testimony to the development of one of the world's oldest writing systems, ancient beliefs and social systems.

6

7

_8 Unearthed jade article of Yin Ruins
_9 Unearthed jade article of Yin Ruins
_10 Oracle bone of Yin Ruins

10

Kaiping Diaolou and Villages

Kaiping, a county situated in the Xijiang River Valley of Guangdong Province, is well known for its Diaolou and villages in the world. In the villages of the county, the number of Diaolou now reaches 1,833. Of them, the Sanmenli Village of Chikan Town, the Zili Village and Fang Clan Watch Tower at Tangkou Town, the Jinjiangli Village at Xiangang Town and Maxianglong Village Cluster at Baihe Town are nominated as the four sites of Diaolou and villages for the World Heritage.

Since the Ming Dynasty, Kaiping County had frequently suffered from floods and banditry, so that some local villagers started to build such Diaolou to defend themselves against such scourges. During the middle of the 19th century, along with the fast development of the United States and western Canada, and the gold rush and railway construction which were in need of enormous imported labors, in addition to the worsened conditions for existence of the local people because of frequent belligerences among themselves, a great number of local farmers began immigrating to America, resulting in many overseas Kaiping workers in the U.S. and Canada. During the 1880s, the two countries followed a policy to squeeze out Chinese workers, so a lot of them were forced to return to their native towns to purchase land, build houses, and get married. From the late 19th century to the 1920s, along with the rapid economic development in the United States and Canada, the Chinese workers' income and remittances of dollars into China also increased. It paved way for the mushrooming of building Diaolou in Kaiping. According to statistics, during the period from 1900 to 1931, a total of 1,648 Diaolou were built, accounting for about 90% of the total.

Kaiping Diaolou is of high cultural value. During the period of the social transition in China at that time, different types of foreign architectural art were earnestly introduced into the rural areas for the construction of the Diaolou, creating an important historical and cultural spectacular in the county and becoming an outstanding example of assimilating foreign culture on a large scale. Meanwhile, they typically reflected the features of the overseas Chinese culture, that is, "taking in everything and making foreign things serve China." For instance, the Ruishi Tower (Ruishi Diaolou), enjoying the fame of "No. 1 Tower of Kaiping," was a steel structure construction with cement inside in western style, using ancient Roman, Byzantine and Baroque architectural elements, as a typical residential Diaolou combining western and Chinese architectural styles. The Liyuan Garden, with its high rising towers, and surrounding vegetations, is honored as one of the typical gardens of overseas Chinese in the county.

On June 28, 2007, Kaiping Diaolou and villages were approved to be included on the list of World Cultural Heritage. The evaluation made by the WHC is as follows:

2

1

3

_1 Ruishi Tower at Kaiping
_2 Mingshi Tower at Kaiping
_3 Liyuan Garden at Kaiping

Kaiping Diaolou and Villages feature the Diaolou, multi-storyed defensive village houses in Kaiping, which display a complex and flamboyant fusion of Chinese and Western structural and decorative forms. They reflect the significant role of émigré Kaiping people in the development of several countries in South Asia, Australasia and North America, during the late 19th and early 20th centuries. There are four groups of Diaolou and twenty of the most symbolic ones are inscribed on the List. These buildings take three forms: communal towers built by several families and used as temporary refuge, residential towers built by individual rich families and used as fortified residences, and watch towers. Built of stone, pise, brick or concrete, these buildings represent a complex and confident fusion between Chinese and Western architectural styles. Retaining a harmonious relationship with the surrounding landscape, the Diaolou testify to the final flowering of local building traditions that started in the Ming period in response to local banditry.

_4 Inside view of Liyuan Garden at Kaiping
_5 Inside view of Liyuan Garden at Kaiping
_6 A western style building of Liyuan Garden at Kaiping

洋

7

8

9

_7 Inside view of Liyuan Garden at Kaiping
_8 Inside view of a room at Zili Village of Kaiping
_9 Inside view of a room at Zili Village of Kaiping
_10 A full view of Liyuan Garden at Kaiping

Fujian *Tulou*

Fujian *Tulou* includes those buildings in the southern part of Fujian Province and Hakkas area, totaling about 3,000. Generally speaking, they are unique large buildings constructed with rammed earth as solid walls for dwelling and defense against any possible attacks from outside. These tulou, or earthen buildings, in a shape of a flying saucer from outer space, are scattered in the fields and hillsides of the mountainous southwest Fujian and the boundary area of Fujian, Guangdong and Jiangxi provinces. Such buildings are gems of traditional Chinese dwellings and the architectural techniques are designated as the national intangible cultural heritages.

On the list of World Cultural Heritage, 46 buildings of the Fujian *Tulou* can be classified into six clusters in addition to four buildings, including Chuxi, Tianluokeng, Hekeng, Hongkeng, Gaobei, Dadi tulou clusters, as well as the Yanxiang Building, Zhenfu Building, Huaiyuan Building and Hegui Building. They are mainly built in a circular shape and some in a square form.

The tulou buildings were first built during the period of Song and Yuan dynasties. They developed through the Ming and Qing dynasties and reached its mature stage in the period of the Republic of China, and lasted till the present time. Their shapes can be divided into circular, square and five-phoenix, deformed with a dent section, semi-circular and the eight-diagramed ones. The origin of the buildings can be traced up to the Tang Dynasty when Chen Yuanguang, an army commander, stationed his troops at Zhangzhou, and 58 households were also settled in southern Fujian. Vestiges of ancient military barracks can be found nowadays on the local round mountain tops. Because of the round top of mountains, no square forms of barracks could be built at that time. These round barracks were the prototype of the tulou buildings. In ancient times, the round barracks had no floors. Later, when they were turned to be used as dwellings for common people, in order to save the farmland, they developed to become buildings of several storeys. Furthermore, when the Hakkas people first moved to Fujian from central China, they were used to their former customs to live in a courtyard dwelling with walls on four sides. But for the purpose of protecting themselves from outside attacks, they constructed the thick-earthen houses, forming the square shaped tulou and the five-phoenix building. By adopting a circular form, the building could have a better distant view from the higher floors, then were accordingly favored by the Hakkas people.

These large earthen dwelling constructions in mountainous area are rarely found in the world. They are noted in the world for their mystic environment for settlement, styles of construction, a cunningly designed defensive system, ingenious craftsmanship, as well as the profound local tulou culture. They were distributed in line with the contour of the terrain, identical with the traditional idea of geomancy for house building, and meeting demands for compact inhabitancy and defensive functions. Making good use

2

_1 The Chuxi tulou cluster
_2 The Tianluokeng tulou cluster
_3 Inside view of a tulou building

_4 Inside view of a tulou building
_5 Inside view of a tulou building
_6 Inside view of a tulou building

of the local building materials, all these buildings were constructed as a singular and independent system with a good aesthetic taste.

In July 2008, at the 32nd session of the World Heritage Commission held in Quebec City of Canada, Fujian *Tulou* was listed as one of the sites of World Cultural Heritage. The evaluation made by the World Heritage Commission is:

Fujian Tulou is a property of 46 buildings constructed between the 15th and 20th centuries over 120 km in south-west of Fujian province, inland from the Taiwan Strait. Set amongst rice, tea and tobacco fields the Tulou are earthen houses. Several storeys high, they are built along an inward-looking, circular or square floor plan as housing for up to 800 people each. They were built for defence purposes around a central open courtyard with only one entrance and windows to the outside only above the first floor. Housing a whole clan, the houses functioned as village units and were known as "a little

5

6

kingdom for the family" or "bustling small city." They feature tall fortified mud walls capped by tiled roofs with wide over-hanging eaves. The most elaborate structures date back to the 17th and 18th centuries. The buildings were divided vertically between families with each disposing of two or three rooms on each floor. In contrast with their plain exterior, the inside of the tulou were built for comfort and were often highly decorated. They are inscribed as exceptional examples of a building tradition and function exemplifying a particular type of communal living and defensive organization, and, in terms of their harmonious relationship with their environment, an outstanding example of human settlement.

7

_7 Inside view of a tulou building
_8 Inside view of a tulou building
_9 The Yanxiang Building
_10 Landscape of tulou buildings

8

衍香楼

Mount Wutai

Mount Wutai is one of the world's five Buddhist sacred lands in the world. It is also the first of China's four famous Buddhist mountains, or one of China Top Ten Mountains. It is one in the first group of the state-level key scenic spots. And it is a national forest park.

Mount Wutai is part of the Taihang mountain range, located in the northeast of Wutai County in Shanxi Province, always known as the "Roof Ridge of North China." Its five peaks, imposing and magnificent, meandering and surrounded each other, have their different charms and special appeals. It is about 250 km in circumference with a total area of 2,837 km^2. Its top is flat and wide, like a piled up terrace. The Peak Overlooking the Sea on the eastern tableland, 2,795m above sea level, is facing the bright sun rise in the east; the Spendor Peak with an elevation of 2,485m on the southern tableland, is strewn all over with grasses and flowers like a blanket of brocade; the Moon-hanging Peak with an elevation of 2,773m on the western tableland, looks as if the moon was suspended above the peak; the Peak of Flourishing Leaves of 3,061m above sea level on the northern tableland, is surrounded by clouds in its waist, and its top looks like to touching the Big Dipper; the Emerald Rock Peak of 2,894m above sea level, with emerald rocks and green cliffs, is the center of the five tablelands. During the mid-summer, it is covered with all kinds of wild flowers which are matched with the age-old solid ice at the north of its top to present a fascinating spectacle.

Since the 11th year of the Yongping period of the Eastern Han Dynasty (68 A.D.) when temples were first built on the mount, about 2,000 years have elapsed. It is the exclusive mountain to hold the temples of both the Chinese Buddhism in Han area and the Yellow Hat Sect of Lamaism in China. All the Buddhists of various nationalities in the country are sincerely paying their utmost homage to the mount. The Emperor Xiaowen of Northern Wei Dynasty, the Emperor Yang of Sui Dynasty, the Emperor Taizong of Song Dynasty, Emperor Yingzong of Yuan Dynasty, and Emperors Shengzu and Gaozong of Qing Dynasty, all had visited the mount in person, and through the ages, all the emperors of every dynasty had never ceased to dispatch their envoys or send messages or gifts to the mount. In the mount, a total of 47 temple constructions of various dynasties from Northern Wei to Tang, Song, Yuan, Ming, and Qing are still kept intact, and the colorful sculptures of seven dynasties, frescos of five dynasties and other architectural arts of ancient times were collected and stored, which are honored as the Buddha's glory, or national treasures of China. Of them, the South Meditation Temple is the oldest construction with wood framework still existing now in the world. Its Temple of Buddha Glory is honored as "a bright pearl of the oriental antique construction." The Big White Pagoda in the Pagoda Courtyard Temple is the existing highest overturned-bowl pagodo in China built in Yuan Dynasty.

The landscape on the mount is magnificent and strange and its humanity cultural sceneries are colorful and diversified. With their appeals and taste of old times, they attract so many Buddhist followers and tourists to come there to pay their worship and for sightseeing. In the mount, a number of senior or master monks have been fostered. During the prime stage of Tang Dynasty, the mount became a Buddhist academy where many overseas Buddhists came to study the Buddhist scriptures and doctrines. Through the ages, a lot of Buddhists coming from India, Japan, Mongolia, Korea, Nepal, and Sri Lanka came to the mount to make a pilgrimage and learn satra. Some of them even remained to stay in the mount to mortify them in a lifelong time.

At the 33rd session of the World Heritage Commission held in Seville, Spain on 26th of June in 2009, Mount Wutai was officially included on the world heritage list. The excerpt of the evaluation of WHC is as follows:

Mount Wutai with its five flat peaks is one of the four sacred Buddhist Mountains in China. It is seen as the global centre for Buddhist Manjusri worship. Its fifty-three monasteries include the East Main Hall of Foguang Temple, with life size clay sculptures, the highest ranking timber building to survive from the Tang Dynasty, and the Ming Dynasty Shuxiang Temple with a huge complex of 500 'suspension' statues, representing Buddhist stories woven into three dimensional pictures of mountains and water. The beauty of its landscape has been celebrated by artists since at least the Tang Dynasty. For a thousand years from the Northern Wei period (471-499) nine Emperors made 18 pilgrimages to pay tribute to the bodhisattvas, commemorated in stele and inscriptions. Started by the Emperors, the tradition of pilgrimage to the five peaks is still very much alive. With the extensive library of books, the monasteries of Mount Wutai remain an important repository of Buddhist culture, and attract pilgrims from across a wide part of Asia.

_1 Image of Manjusri Bodhisattva on Mount Wutai
_2 Stone archway at Longquan Temple of Mt. Wutai
_3 Big White Pagoda on Mt. Wutai
_4 The Taihui Town of Mt. Wutai

4

Mixed World Cultural and Natural Heritage

Mount Taishan

Mount Taishan is located in the central part of Shandong Province, adjoining Jinan in the north, facing Qufu in the south, bordering on Zibo in the east and close to the Yellow River in the west, stretching over 426 km^2. The Jade Emperor Peak, its main peak, is 1,545 m above sea level, imposing and magnificent.

In ancient times Mount Taishan was called Dai Ancestor, and renamed as Taishan during the Spring and Autumn Period (770 B.C.-476 B.C.). Tai in Chinese literally means extremely large, easiness and peace. Towering over the ancient states of Qi and Lu, on the North China Plain, Mount Taishan together with Mt.Hengshan (in Hunan), Mt. Hengshan (in Shanxi), Mt. Huashan, and Mt. Songshan are known as the "Five Sacred Mountains" in China. Mt. Taishan is also known as the East Sacred Mountain. Through the ages, Mt. Taishan had been worshipped as a divine mountain for emperors of all the dynasties to pay homage to Heaven and for deep meditation. Many Taoist priests, Buddhist monks and other celebrities were inspired to visit the mountain and left over a lot of cultural relics, making it a great symbol of oriental civilization.

Centering on the main peak, the scenic areas of the mountain are distributed in all directions, involving both the natural landscape and human and cultural sceneries. Starting from the Dai Temple to the Jade Emperor Peak, a 10 km-long passage has been formed for the tourists to visit scenic spots along the route of "the world—the hell—the paradise." With its grand and sublime mountains, strewn with pines and cypresses, landscape in Mt. Taishan boasts a quality of primitiveness, splendor, seclusion and wonders. Steles on cliffs with inscriptions by literates and celebrities through the ages can be encountered anywhere. Of the scenic spots, the most famous ones include the Heaven Pillar Peak, the Sunrise Watching Peak, the Baizhang Cliff, the Immortal Bridge, Pine of Five Grand Masters, the Expectation Pine, as well as the landscapes of the Dragon Pool with Flying Waterfall, the Cloud Bridge with Flying Waterfall, and the Three Pools with Flying Waterfalls. In the mountains, there are about 20 antique construction complexes and about 2,000 cultural remains. It has four well known wonderful spectaculars, including that of Sunrise Over the Mountain, the Sea of Clouds, Glow of a Setting Sun and the Golden Belt along the Yellow River.

In December 1987, according to the standards for selection of the mixed cultural and natural heritage C (I) (II) (III) (IV) (V) (VI) and N (III), Mt. Taishan was listed as one of the sites of World Heritage. The evaluation made by the World Heritage Commission is as follows:

The sacred Mount Tai ('shan' means 'mountain') was the object of an imperial cult for nearly 2,000 years, and the artistic masterpieces found there are in perfect harmony with the natural landscape. It has always been a source of inspiration for Chinese artists and scholars and symbolizes ancient Chinese civilizations and beliefs.

2

_1 The Jade Emperor Peak of Mount Taishan
_2 The First Gate to Heaven at Mount Taishan
_3 The Hall of Heavenly Grant in the Dai Temple at Mount Taishan

_4 A memorial stone archway of Mt. Taishan
_5 Stone inscription complex known as “A Grand View of the World” at Mt. Taishan
_6 Eighteen Bends at Mt. Taishan
_7 The Hall of Heavenly Grant in the Dai Temple at Mount Taishan

_8 A memorial stone archway of Mt. Taishan
_9 The sacred divine image of Taishan in the Dai Temple at Mt. Taishan
_10 The Azure Cloud Temple at Mt. Taishan

東嶽泰山之神

Mount Huangshan

Called "Mount Yishan" in ancient times, Mt. Huangshan had long since been known as a Taoist sacred mountain, 250 km in circumference. It is said that the Yellow Emperor (a legendary ruler in primitive society) had been there mortifying himself and practicing the alchemy, and later he attained to the enlightenments to ascend to Heaven. Hence Emperor Xuanzong of Tang Dynasty sent edict to name it Huangshan in the sixth year of the Tianbao period (747 A.D.).

Mt. Huangshan is noted for its magnificent, grotesque, dream-like and precipitous landscape, and is one of the best known scenic areas in China. Strewn with delicate and queer-looking rocks in various shapes and changing landscapes in different seasons, it looks as if it were a fairyland under the heaven. It has 77 peaks each above 1,000 m in elevation, 36 large peaks and 36 small ones. Its three main peaks, Lotus Peak, Heavenly Capital Peak, and Bright Peak are all 1,800 m above sea level. The noted traveler Xu Xiake of the Ming Dynasty lauded the mountain: "People need not to visit any other mountains when they have toured the Five Sacred Mountains; and need not to visit any other sacred mountains when they have toured Mt. Huangshan."

The beauty of Mt. Huangshan lies in its "five wonders": legendary pines, grotesque rocks, sea of clouds, hot springs, and snow in winter. It boasts abundant resources of biodiversity. It is densely covered by trees of which some ancient ones tower up to the sky. It is also a habitat of so many rare beasts and birds. About 1,500 species of plants and more than 500 species of animals have been found there. The Huangshan pines are the most attractive. The noted rock sceneries include Golden Pheasant Calling at the Heavenly Gate, Squirrels Jumping at Heavenly Capital, and Stone Monkey Watching the Sea of Clouds. The mountain is often enveloped by a sea of clouds, appearing in all its majesty. Its hot spring, or divine spring, originates from beneath the Purple Cloud Peak which is 850 m above sea level, and never dries up even in a lasting drought season. Its limpid water mainly contains heavy carbonate and other microelements, keeping a constant temperature of 42℃. The water is potable or can be used for bathing. The mountain has a history replete with rich historical and cultural anecdotes and was the birthplace of the traditional Chinese Shanshui "mountain and water" style of painting. There are some 200 temples, pavilions, spiral roads, ancient bridges, and cliff carving calligraphy, being left over to the present days through the ages. Its snow scenery in winter will become a spectacular when it drops down upon pines and rocks and meets with springs and clouds, a wonderful picture of flying snow, the hanging icicles, and rimed pines.

In 1982, the mountain was designated

2

_1 A scene of the mist and cloud at Mount Huangshan
_2 The Flying Stone at Mount Huangshan
_3 The Guest-greeting Pine at Mount Huangshan

by the State Council as one in the first group of the state-level key scenic spots, ranking in parallel with the Yellow River, Yangtze River and the Great Wall in terms of their cultural significance and value. In December 1990, it was included on the list of the Mixed World Cultural and Natural Heritage by UNESCO. The evaluation made by the WHC is:

Huangshan, known as 'the loveliest mountain of China', was acclaimed through art and literature during a good part of Chinese history (e.g. the Shanshui 'mountain and water' style of the mid-16th century). Today it holds the same fascination for visitors, poets, painters and photographers who come on pilgrimage to the site, which is renowned for its magnificent scenery made up of many granite peaks and rocks emerging out of a sea of clouds.

_4 The scene of rimed pines at Mount Huangshan

5

6

_5 Dawn at Mount Huangshan
_6 The Heavenly Capital Peak at Mount Huangshan
_7 The beauty of peaks at Mount Huangshan
_8 A sea of clouds at Mount Huangshan

Mount Emei and Leshan Giant Buddha

Mt. Emei is situated in the south-central part of Sichuan Province, and the highest Ten Thousand Buddha Top of its principal Golden Peak is 3,099 m above sea level. Its beautiful landscape together with its mythic Buddhist legend on the divine mountain is well known in the world. It is one of China's noted mountains, for its magnificent, graceful, mythic, and strange landscape.

Mt.Emei is the Buddhist rite of Samantabhadra Bodhisattva. It is also one of the four famous Buddhist Mountains in China. The popularization of Buddhism as well as the construction and prosperity of Buddhist temples in China have brought about a mystic luster to the mountain, producing a special cultural accumulation. Numerous temples stand on the mountain, of which the most famous ones include the "Eight Large Temples on the Golden Peak," such as the Baoguo Temple, and the Wannian Temple.

The Leshan Giant Buddha, sitting at the Perching Phoenix Peak on the east side of Mt. Emei, was first cut in the first year of Kaiyuan period of Tang Dynasty (713), and the work continued until after 90 years. The Giant Buddha is a sitting statue of Maitreya located at the mountain side and close to the river, facing to the west, looking dignified in visage. It is 71 m tall, the existing largest Cliff stone statue in the world. Exquisitely cut, easily shaped, and well proportioned, the Buddha suggests an imposing air, embodying the verve of the culture and art of the highly developed Tang Dynasty. On the cliffs of the north and south sides of the statue, about 90 niches of elegantly cut Buddha statues were placed.

Mt. Emei is situated in an area where many natural elements meet together. It boasts abundant biological species, and many of special species and strains. And the subtropical vegetations in the mountain are well preserved, including 3,200 species in 242 families of valuable plants, accounting for about one-tenth of the country's total. It is also a habitat of many species of rare animals. More than 2,300 species of animals have been found in the mountain, making it an important realm for the study of world biota.

In December 1996, according to the standards of site selection C (IV)(VI), N(IV), UNESCO included Mt. Emei and Leshan Giant Buddha on the list of Mixed Cultural and Natural Heritage. The evaluation made by the World Heritage Commission is as follows:

The first Buddhist temple in China was built here in Sichuan Province in the 1st century A.D. in the beautiful surroundings of the summit Mount Emei. The addition of other temples turned the site into one of Buddhism's holiest sites. Over the centuries, the cultural treasures grew in number. The most remarkable is the Giant Buddha of Leshan, carved out of a hillside in the 8th century and looking down on the confluence of three rivers. At 71 m high, it is the largest Buddha in the world. Mount Emei is also notable for its exceptionally diverse vegetation, ranging from subtropical to subalpine pine forests. Some of the trees there are more than 1,000 years old.

2

_1 The Golden Peak of Mount Emei
_2 A Temple of Mount Emei
_3 The Wannian Temple at Mount Emei
_4 Snowscape of Mount Emei

_5 The snow-capped Golden Peak of Mount Emei
_6 A statue of Buddha of Mount Emei
_7 Sunrise at the Golden Peak of Mount Emei

_8 Leshan Giant Buddha
_9 A hand of Leshan Giant Buddha
_10 A foot of Leshan Giant Buddha
_11 A full view of Leshan Giant Buddha

Mount Wuyi

Mt.Wuyi makes up a total area of 999.75 km^2, situated about 15 km south of the former Chong'an County in the northwestern part of Fújian Province. Its main scenic area covers 70km^2. Mt. Wuyi is of the typical Danxia landform, with an average 350 m in elevation, always honored as a land of "green water and red mountains," and "a land of unique elegance in southeast China." It is one in the first group of the state-level key scenic spots. From time immemorial, the power of nature has created all its strange peaks, green waters and red mountains, producing the beautiful landscape of Mt. Wuyi.

In its scenic area, there are 36 peaks, 72 caves, 99 crags, and 108 scenic spots, distributed in seven scenic sections: Wuyi Palace, Nine Bend River, the Peach Blossom Cave, Heavenly Tour Peak, Sight of the line-like sky---Tiger Roaring Peak, Crag of Heaven Center, and the Water-Curtain Cave. Its first peak is the Great King Peak, in the shape of a black gauze cap, towering up like a heavenly pillar. The vegetations in the mountain are well preserved, and rich biological resources have been found. The beauty of the mountain was appraised by the ancient people as strange as Mt.Huangshan, elegant as Guilin, majestic as Mt. Taishan, precipitous as Mt. Huashan and graceful as the West Lake.

The vivacity of the mountain lies in its waters. The most beautiful is the Nine Bend River, extending 62.8 km long, from west to east, winding around the bottoms of mountains. Sailing on the old-fashioned bamboo raft along the river, visitors would have a feeling as if they came into a mystic fairyland. Moreover, visitors will be surprised to find those boat-shaped coffins inserted and suspended on the high cliffs about 3,800 years ago.

Mount Wuyi is also a mountain of historic and cultural significance, strewn with vestiges of ancient literates. They eulogized it in essays and poems and have left over about 2,000 poems to the present time. Of them, the most famous writer was Zhu Xi, the celebrated scholar of the Neo-Confucianism of Southern Song Dynasty. He had lived in the mountain with his mother for many years and invented the noted Wuyi Eight–Diagrammed Banquet and left over much philosophical thinkings. In the mountain, there are about 400 cliff stone inscriptions left over by ancient people. Also found in the mountain were the ruins of a Han period city, an ancient porcelain kiln of Song Dynasty and an imperial tea garden of Yuan Dynasty.

On December 1, 1999, at the 23rd session of the World Heritage Commission of UNESCO, Mt. Wuyi was listed as one site of the Mixed Cultural and Natural Heritage. The evaluation made by the WHC is as follows:

Mount Wuyi is the most outstanding area for biodiversity conservation in south-east China and a refuge for a large number of ancient, relict species, many of them endemic to China. The serene beauty of the dramatic gorges of the Nine Bend River, with its numerous temples and monasteries, many now in ruins, provided the setting for the development and spread of neo-Confucianism, which has been influential in the cultures of East Asia since the 11th century. In the 1st century B.C. a large administrative capital was built at nearby Chengcun by the Han dynasty rulers. Its massive walls enclose an archaeological site of great significance.

1

3

_1 Mount Wuyi surrounded by mountains and waters
_2 The picturesque Mount Wuyi
_3 Jade Lady Peak looking from the Nine Bend River
_4 Sailing on the Nine Bend River
_5 Mount Wuyi enveloped by mists

World Natural Heritage

Wulingyuan Scenic and Historic Interest Area

Situated in the Zhangjiajie City, in the northwestern part of Hunan Province, and covering a total area of 264 km^2, the Wulingyuan scenic and historic interest area includes the three best-known scenic spots of Zhangjiajie National Forest Park, Suoxi Valley and Tianzi Mountian, and has a landform of peaks formed with quartz sandstone.

It is noted in the world for its strange peaks, grotesque rocks, sequestered valleys, limpid streams and karst caves, known as the "five unrivaled beauties." With its typical landform structure formed through the corrosion and denudation and accumulations of valley erosion, the area boasts so many gullies, streams and brooks in diverse shapes. In the area, there are 3,103 marvelous peaks, and a great number of underground karst caves, the best-known of which is the Yellow Dragon Cave at Suoxi Valley, extending 7.5 km long with a vast interior space reaching four-storey high. Its ecosystem, full of various biological resources, is well preserved with a high coverage rate of forest up to 85%. It has more than 3,000 species of dominating plants, about 700 arbor species, 450 kinds of ornamental flowers and 116 species in 50 families of terrestrial vertebrate.

In December of 1992, Wulingyuan was included on the list of World Heritage according to the selection standards N (III). The evaluation made by the World Heritage Commission is:

A spectacular area stretching over more than 26,000 hm2 in China's Hunan Province, the site is dominated by more than 3,000 narrow sandstone pillars and peaks, many over 200 m high. Between the peaks lie ravines and gorges with streams, pools and waterfalls, some 40 caves, and two large natural bridges. In addition to the striking beauty of the landscape, the region is also noted for the fact that it is home to a number of endangered plant and animal species.

2

3

_1 Spectacular sandstone peaks of Wulingyuan
_2 Beautiful landscape of Wulingyuan
_3 Sheer sandstone pillars of Wulingyuan
_4 Grotesque peaks of Wulingyuan

_5 Grotesque peaks of Wulingyuan
_6 Dawn at Wulingyuan
_7 The Shentang Bend of Wulingyuan

8

_8 A distant view of Wulingyuan
_9 A picturesque scene of Wulingyuan
_10 The Xihai Stone Forest of Wulingyuan

9

10

Jiuzhaigou Valley Scenic and Historic Interest Area

Jiuzhaigou Valley is located in Ngawa, northwest of Sichuan Province, against the northern side of the Galna Peak on the southeast verge of the Qinghai-Tibet Plateau, stretching over about 720 km^2. Most of the area is dominated by forests and lakes. Since there are nine Tibetan villages nestling in the valley surrounded by high mountains and lakes, hence the name Jiuzhaigou.

Jiuzhaigou with its beautiful landscape, has been known as the "fairyland," or "a wonderland on earth." Its major interesting places consist of two gullies including "Goukou—Nuorilang—Changhai" and "Nuorilang—primitive forest" to form a "Y" shaped avenue, extending more than 50 km long. These gullies are dotted with many lakes, waterfalls and calcareous tufa shoals, creating the six wonderful spectaculars: Changhai, Sword Rock, Nuorilang, Shuzheng, Zharu and Heihai. Of all the sceneries, the most attractive is the charms of waters. The crystal-like and clear brooks and streams crisscross through the forests and lakes just like many strings of necklaces. The 114 colorful and limpid lakes together with numerous springs, waterfalls, rivers and shoals in the area, are linking up to present a beautiful painting or a lyric poem. Moreover, with the snow-capped peaks under the azure sky in the changeable four seasons, the landscape of Jiuzhaigou enjoys the honor of "the best water scenery of China."

In December 1992, Jiuzhaigou was listed as one of the World Natural Heritage sites by the World Heritage Commission (WHC) according to the selection standards of N (III). The evaluation made by the WHC is:

Stretching over 72,000 hm2 in the northern part of Sichuan Province, the jagged Jiuzhaigou valley reaches a height of more than 4,800 m, thus comprising a series of diverse forest ecosystems. Its superb landscapes are particularly interesting for their series of narrow conic karst land forms and spectacular waterfalls. Some 140 bird species also inhabit the valley, as well as a number of endangered plant and animal species, including the giant panda and the Sichuan takin.

2

3

1

4

_1 Brooks at Jiuzhaigou Valley
_2 A green lake at Jiuzhaigou Valley
_3 Landscape of Jiuzhaigou Valley
_4 A water-operated mill of Jiuzhaigou Valley

_5 Landscape of Jiuzhaigou Valley
_6 Waterfall of Jiuzhaigou Valley

6

_7 Landscape of Jiuzhaigou Valley
_8 Landscape of Jiuzhaigou Valley
_9 A green lake of Jiuzhaigou Valley

Huanglong Scenic and Historic Interest Area

Huanglong Scenic and Historic Interest Area is located at the Songpan County of the Ngawa Tibetan and Qiang Autonomous Prefecture, covering an area of about 700 km^2. with an average elevation of above 3,000 m and consisting of the Huanglong and Mouni Valley scenic areas. It is famed in the world for its unique karst sceneries—surface travertine and abundant fauna and flora resources. With its magnificent, precipitous, strange and wild landscapes, the area has been honored as a "fairyland on earth."

Its main scenic area is the 3.6-km-long Huanglong Valley, which is widely covered by deposits of carbonate calcareous tufa, in tiers of terraced fields, looking like a giant golden dragon, amidst which various sceneries of delicate lakes, pools, shoals, waterfalls, springs and karst caves are randomly dotted. A majestic scene would appear when they are illuminated with the presentation of snow-clad mountains, the primitive forest, valleys and ravines. From the bottom of the Huanglong Valley (2,000 m above sea level) to the top of it (3,800 m above sea

level), the landscapes of the mixed forests of subtropical evergreen and broadleaf and abscission plants, mixed forest of coniferous and broadleaf plants, high-mountain coniferous forest and scrubs and meadows are gradually laid out one after another. It is the habitat of over ten species of precious and rare animals including giant panda and golden monkey.

In December 1992, the Huanglong Scenic and Historic Interest Area was included on the World Heritage site list. The evaluation made by the World Heritage Commission is:

Situated in the north-west of Sichaun Province, the Huanglong valley is made up of snow-capped peaks and the easternmost of all the Chinese glaciers. In addition to its mountain landscape, diverse forest ecosystems can be found, as well as spectacular limestone formations, waterfalls and hot springs. The area also has a population of endangered animals including the giant panda and the Sichuan golden snub-nosed monkey.

1

_1 Huanglong at dusk
_2 A colorful pond of Huanglong
_3 A waterfall scene of Huanglong

_4 A verdant landscape of Huanglong

_5 Huanglong in winter

6

_6 Beautiful landscape of Huanglong
_7 The Multi-color Pond of Huanglong

Three Parallel Rivers of Yunnan Protected Areas

"Three rivers run in parallel with each other" is referred to a bizarre natural geographical scenery created by the Jinsha River, the Lancang River and the Nujiang River, all originating from the Qinghai-Tibet Plateau, that run in parallel with each other about 170 km long in Yunnan Province from north to south without confluence, through many towering mountains such as Mt.Dandanglica, Gaoligong Mountains, Mt. Nushan, and Mount Yunling. They appear at the joint point of the three geographical bounds of the East Asia, South Asia and the Qinghai-Tibet Plateau, where is the typical region of high mountainous landform and its evolution, and one of the biodiversity regions in the world.

The area's unique geological structure, a center of biodiversity in the Eurasia, its rich humanistic and cultural resources, and beautiful natural landscapes make up its unique wonders in the world. Meanwhile, the area is the compact communities of 16 ethnic minority groups of China. Despite coming from different ethnic groups, using different languages and believing in different religions, they are all leading a harmonious life with each other. The Shangri-La, a supposed sequestered land of peace and happiness under the sun, a household word in the western countries, would surely find its origin in there. In the 1920s, an American writer Joseph Rock, had written stories and taken pictures for the magazine National Geographic and collected various species from the area for almost 20 years. The *Lost Horizon*, a book written by an Englishman James Hilton, introduced the charm of the landscapes in this "Shangri-La" to the whole world. The well-known international trade passage in China—the Tea-horse Ancient Road also originated from the area.

In July 2003, the Yunnan Protected Areas were included on the list of World Natural Heritage sites according to the selection standards of N(I) (II) (III) and (IV). The evaluation made by the WHC is:

Consisting of eight geographical clusters of protected areas within the boundaries of the Three Parallel Rivers National Park, in the mountainous north-west of Yunnan Province, the 1.7 million hectare site features sections of the upper reaches of three of the great rivers of Asia: the Yangtze (Jinsha), Mekong and Salween run roughly parallel, north to south, through steep gorges which, in places, are 3,000 m deep and are bordered by glaciated peaks more than 6,000 m high. The site is an epicentre of Chinese biodiversity. It is also one of the richest temperate regions of the world in terms of biodiversity.

The sites included on the World Natural Heritage list were nominated by eight geographical clusters. The Nujiang River, the Lancang River, the Jinsha River, amidst four mountains, link up the eight clusters as a whole. Each of them has become a "puzzled question" to the world, representing respectively specific diversity of species, geologic structures and sceneries under different geographical conditions. They compensate one another in terms of the values of themselves, while they are irreplaceable in terms of their typical natural resources. As a result, they altogether have produced "exclusion and integrity" of the value of geological resources as a world natural heritage site.

_1 The Nujiang River in Yunnan Province

2

_2 The Songzanlin Monastery at Zhongxun County in Yunnan Province
_3 The Tiger Leaping Gorge of Jinsha River in Yunnan Province

_4 The First Bend of the Yangtze River at Shigu Town in Yunnan Province
_5 The Mount Siguniang at Ngawa in Sichuan Province

Sichuan Giant Panda Sanctuaries

_1 Wolong in winter

Giant Panda is one of the world's most precious animals on the verge of extinction. It is known as "the living fossil." By the end of 2005, about 180 giant pandas had lived in captivity in the whole world. And a total of 1,590 wild pandas survive in the world, mainly distributed in Sichuan and Shaanxi provinces in China. In Sichuan Province, their habitats include Wolong, Mt. Siguniang and Jiajin Mountains, covering a total area of 9,245 km^2, across 12 counties under the jurisdiction of four cities or prefectures of Chengdu, Ya'an, Ngawa and Ganzi, and organized by seven natural reserves and nine scenic spots, where about 30% of the total world's pandas inhabit. Here are the most complete and the largest giant panda habitats in the world. Moreover, they are one of the regions abundant in most vegetation species in the world except tropical rain forest areas.

The environment of panda's sanctuaries in Sichuan Province is similar to that of the paleo-tropic forests of the Tertiary Period dating back to several million years. It is one of the 25 hot spots of biodiversity selected by the Conservation International. In the area, there are more than 10,000 kinds of advanced plants and rare animals such as giant panda, golden monkey, golden takin, red panda, snow leopard, and clouded leopard, well known as the "biotic museum" of the world. So, it has also been designated as one of the 200 ecological areas by the WWF.

On the 12th of July 2006, a decision was unanimously adopted by the 30th session of the World Heritage Commission of UNESCO, that the panda sanctuaries in Sichuan Province were included on the list of the World Natural Heritage. The evaluation made by the WHC is:

Sichuan Giant Panda Sanctuaries, home to more than 30% of the world's pandas which are classed as highly endangered, covers 924,500 ha with seven nature reserves and nine scenic parks in the Qionglai and Jiajin Mountains. The sanctuaries constitute the largest remaining contiguous habitat of the giant panda, a relict from the paleo-tropic forests of the Tertiary Era. It is also the species' most important site for captive breeding. The sanctuaries are home to other globally endangered animals such as the red panda, the snow leopard and clouded leopard. They are among the botanically richest sites of any region in the world outside the tropical rainforests, with between 5,000 and 6,000 species of flora in over 1,000 genera.

_2 A blue brook mirrors the green trees
_3 Verdant mountain upon mountain

South China Karst

Karst is a type of topography developed on carbonate rock composed mainly of dolomite and limestone. China's karst features a large coverage in area, diversified in types, typical and rich in biological ecosystem. The karst in South China accounts for 55 percent of the country's total karst area. It is a project to be submitted in batches and categories by the Chinese government to the World Heritage Commission for inclusion in the list of world natural heritage. The first batch of the project to be listed includes the sword, column and tower karsts in the stone forests in Yunnan Province, and the cone karsts of Libo in Guizhou Province, and the Wulong karst represented by natural bridges, caves and giant dolines in Chongqing Municipality. They had developed and been formed in a period dating back to half a million or 300 million years, covering a total area of 1,460 km^2. Its core area covers 480 km^2. and its buffer area 980 km^2.

South China karst is the most typical in China, where most of karst landforms are concentrated, and many of them are well known in the world. For instance, the stone forests in Yunnan Province is famed for its magnificent, strange, forbidding, elegant, secluded, curious and wild beauties and is honored as the "cream of karst of the world." The Libo karst in Guizhou, which is the inhabitant community of many ethnic minority groups such as Buyi, Shui, Miao and Yao, boasts its typical scenery—Maolan karst forest. In 2005, when the public appraisal activity was held by the magazine China National Geography for "China's Most Beautiful Place," this forest was selected as the most beautiful forest in China. According to the types of karst combining with the types of forest, Lipo karst primitive forest sceneries can be further classified into four categories: funnel type, depression type, basin type and gully type forest. The Wulong karst is unique in the world, including "the first cave under the sun" Furong Cave, the largest group of natural bridges in Asia and the giant dolines at Houping which are all rare wonders.

On the 27th of June, 2007, at the 31st session of the UNESCO World Heritage Commission, a decision was unanimously adopted to name South China Karst as one of the sites of World Natural Heritage. The evaluation made by the WHC is:

The South China Karst region extends over a surface of half a million km2 lying mainly in Yunnan, Guizhou and Guangxi provinces. It represents one of the world's most spectacular examples of humid tropical to subtropical karst landscapes. The stone forests of Shilin are considered superlative natural phenomena and a world reference with a wider range of pinnacle shapes than other karst landscapes with pinnacles, and a higher diversity of shapes and changing colours. The cone and tower karsts of Libo, also considered the world reference site for these types of karst, form a distinctive and beautiful landscape. Wulong Karst has been inscribed for its giant dolines (sinkholes), natural bridges and caves.

_1 Stone forests at Guilin of Guangxi Province
_2 A giant doline at Wulong of Chongqing Municipality
_3 A giant doline at Wulong of Chongqing Municipality

1

2

_4 The cone karsts at Libo in Guizhou Province
_5 Stone forests at Guilin of Guangxi Province
_6 Stone forests at Guilin of Guangxi Province

剑峰

Mount Sanqingshan National Park

Situated in the Shangrao City, northeast of Jiangxi Province, the Mount Sanqingshan National Park covers a total area of 229 km^2. Its highest Yujing Peak is rising up to 1,816.9 m high. Since the three peaks Yujing, Yuhua and Yuxu are said to be the shrines of the Taoist supreme trinity deities, called gods of Yuqing, Shangqing and Taiqing, hence the name. Mount Sanqingshan, through the geological movement in a long period of 1.4 billion of years, has formed a series of unique magnificent and grandeur landscapes with queer-looking peaks and grotesque rocks, rapid and high waterfalls, gorges surrounded by quiet-sailing clouds. The whole area is dominated by densely growing ancient trees and luxurious vegetation and is also the habitat of many rare birds and animals.

The mount is well known for its beautiful natural landscape and cultural places of interest, featuring the Taoist style temples and rare fauna and flora. With many wonderful or graceful peaks, the mountain is honored as "the sister mountain of Mt. Huangshan," on which there are ten famous spectaculars such as the Peak of Goddess, Avalokitesvara Bodhisattva Listening to Chinese Lute Playing, Boa Coming Out of the Mountain Cave, Old Taoist Priest Worshipping the Moon, Elegant Peak On a Fairyland, Monkey King Offering Treasures, Monk Tang Going West, Wooden Fish in the Golden Hall, Divine Dragon Playing With a Pine, and Twin Stars in the Chinese Valentine's Day. The great writer Su Dongpo of the Song Dynasty wrote a poem to highly praise the landscapes: "After having admired all the beauties of the Five Sacred Mountains, finally I found the Mount Sanqingshan distinguished itself as the best."

The mount is also a holy mountain of Taoism, with a long history of more than 1,600 years, had been the mortification site of Taoist priests. During the Eastern Jin Dynasty, since Ge Yun and Ge Hong came to the place to mortify and cultivate themselves according to the Taoist doctrines, it had become a sanctuary for many celebrated Taoist priests. The Sanqinggong Monastery as well as 220 other ancient constructions and many stone-carvings are still kept intact, known as the "Museum of Chinese Taoism."

On the 8th of July 2008, at the 32nd session of the UNESCO World Heritage Commission, a decision was adopted that the Mount Sanqingshan was included on the list of the World Natural Heritage and a high evaluation was made on it:

Mount Sanqingshan National Park, a 22,950 hm2 property located in the west of the Huaiyu mountain range in the northeast of Jiangxi Province (in the east of central China) has been inscribed for its exceptional scenic quality, marked by the concentration of fantastically shaped pillars and peaks: 48 granite peaks and 89 granite pillars, many of which resemble human or animal silhouettes. The natural beauty of the 1,817 metre high Mount Huaiyu is further enhanced by the juxtaposition of granite features with the vegetation and particular meteorological conditions which make for an ever-changing and arresting landscape with bright halos on clouds and white rainbows. The area is subject to a combination of subtropical monsoonal and maritime influences and forms an island of temperate forest above the surrounding subtropical landscape. It also features forests and numerous waterfalls, some of them 60 metres in height, lakes and springs.

_1 Mount Sanqingshan enclosed in mists

2

3

4

_2 Winter scene of Mount Sanqingshan
_3 Romantic view of Mount Sanqingshan
_4 A torrent limpid brook
_5 Early spring comes to Mount Sanqingshan
_6 Picturesque scene of Mount Sanqingshan

Chronology of Inclusion of China's Sites on the World Heritage List

Name of Sites	Year of Inclusion
The Great Wall (cultural site)	1987
Imperial Palaces of the Ming and Qing Dynasties in Beijing and Shenyang (cultural site)	1987
Mogao Caves (cultural site)	1987
Mausoleum of the First Qin Emperor (cultural site)	1987
Peking Man Site at Zhoukoudian (cultural site)	1987
Mount Taishan (mixed cultural and natural site)	1987
Mount Huangshan (mixed cultural and natural site)	1990
Wulingyuan Scenic and Historic Interest Area (natural site)	1992
Jiuzhaigou Valley Scenic and Historic Interest Area (natural site)	1992
Huanglong Scenic and Historic Interest Area (natural site)	1992
Mountain Resort and its Outlying Temples, Chengde (cultural site)	1994
Temple and Cemetery of Confucius and the Kong Family Mansion in Qufu (cultural site)	1994
Ancient Building Complex in the Wudang Mountains (cultural site)	1994
Historic Ensemble of the Potala Palace, Lhasa (cultural site)	1994
Lushan National Park (cultural site)	1996
Mount Emei and Leshan Giant Buddha (mixed cultural and natural site)	1996
Classical Gardens of Suzhou (cultural site)	1997
Ancient City of Pingyao (cultural site)	1997
Old Town of Lijiang (cultural site)	1997

Name of Sites	Year of Inclusion
Summer Palace in Beijing (cultural site)	1998
Temple of Heaven in Beijing (cultural site)	1998
Dazu Rock Carvings (cultural site)	1999
Mount Wuyi (mixed cultural and natural site)	1999
Mount Qingcheng and the Dujiangyan Irrigation System (cultural site)	2000
Imperial Tombs of the Ming and Qing Dynasties (cultural site)	2000
Longmen Grottoes (cultural site)	2000
Ancient Villages in Southern Anhui —— Xidi and Hongcun (cultural site)	2000
Yungang Grottoes (cultural site)	2001
Three Parallel Rivers of Yunnan Protected Areas (natural site)	2003
Capital Cities and Tombs of the Ancient Koguryo Kingdom (cultural site)	2004
Historic Centre of Macao (cultural site)	2005
Yin Xu (cultural site)	2006
Sichuan Giant Panda Sanctuaries (natural site)	2006
Kaiping Diaolou and Villages (cultural site)	2007
South China Karst (natural site)	2007
Fujian *Tulou* (cultural site)	2008
Mount Sanqingshan National Park (natural site)	2008
Mount Wutai (cultural site)	2009

English Translator: Li Xingjian

Photographs by: Zhang Zhenguang, Du Yiming, Li Min, etc.

Directed by: Zhang Huizhen
Executive Editors: Zhang Zhenguang, Bai Yumei, Sun Shuyan
Layout Design: Fu Jinhong
Cover Design: Feng Yizheng, Kang Yu

The Chinese Treasure
WORLD HERITAGE IN CHINA

Compiled by Luo Zhewen, Li Min

Published and distributed by China Architecture & Building Press in China
ISBN 978-7-112-12142-7(19464)
CIP data available on request
www.cabp.com.cn

Printed on acid-free paper produced from chlorine-free pulp. TCF∞
Printed in China